"WHY ME?"

"Why me?"—why one person becomes ill while another, also exposed to a disease, does not—is, in fact, the central question confronting medicine today.

The question is important primarily because medical science has done so well at finding out why people *in general* come down with a particular disease; literally hundreds of illnesses, from malaria to pneumonia, have offered up their secrets during the past century. It is central because the diseases that are rampant today—cancer and emphysema and heart problems— seem to strike almost at random, with medical understanding of the reasons often limited to "it runs in the family." It is critical because *preventing* illness, rather than curing it after it strikes, is becoming the watchword of modern medicine.

"Why me?" means so much because if we can determine why one particular individual comes down with an illness, we might also learn to cut the chain of events that causes it, to avoid potential trouble spots in the environment, to know, exactly, who needs what kind of medical advice.

BANTAM NEW AGE BOOKS

This important new imprint—to include books in a variety of fields and disciplines—will deal with the search for meaning, growth and change. BANTAM NEW AGE BOOKS will form connecting patterns to help understand this search as well as mankind's options and models for tomorrow. They are books that circumscribe our times and our future.

Ask your bookseller for the books you have missed

GENETIC PROPHECY: BEYOND THE DOUBLE HELIX

Dr. Zsolt Harsanyi

and

Richard Hutton

BANTAM BOOKS

TORONTO · NEW YORK · LONDON · SYDNEY

This low-priced Bantam Book
has been completely reset in a type face
designed for easy reading, and was printed
from new plates. It contains the complete
text of the original hard-cover edition.
NOT ONE WORD HAS BEEN OMITTED.

GENETIC PROPHECY: BEYOND THE DOUBLE HELIX

A Bantam Book / published by arrangement with
Rawson Wade Publishers Inc.

PRINTING HISTORY

Rawson, Wade edition published July 1981
An Alternate Selection of Book-of-the-Month Club,
Book-of-the-Month Club/Science Main Selection and
Preferred Choice Bookplan Alternate Selection.
Bantam edition / September 1982

New Age and the accompanying figure design
as well as the statement "a search for meaning, growth
and change" are trademarks of Bantam Books, Inc.

ISBN 0-553-22601-0

Published simultaneously in the United States and Canada

Bantam Books are published by Bantam Books, Inc. Its
trademark, consisting of the words "Bantam Books" and
the portrayal of a rooster, is Registered in U.S. Patent
and Trademark Office and in other countries. Marca
Registrada. Bantam Books, Inc., 666 Fifth Avenue, New York,
New York 10103.

PRINTED IN THE UNITED STATES OF AMERICA

O   0 9 8 7 6 5 4 3 2 1

To Harsanyis and Huttons
too numerous to name

AUTHORS' NOTE

In any collaboration, there is a tendency for those outside the process to assign roles to the participants, to assume, for instance, that the scientist did the thinking, the writer did the writing, and that a collaborative caste system was etched in stone. That is not the case with this book. One of us came to this project as a geneticist, the other as a professional observer of the genetic revolution. Roles, as they may be perceived from without, never existed. Every idea, every phrase, every nugget of information was thrown into the communal stew and blended with the rest, until it has become impossible for us—much less for anyone else—to separate our contributions. This book is, in short, a true collaboration; the responsibility for its content rests with both of us, equally.

Our deepest gratitude to those who helped us throughout our work. We wish to thank especially Susan Allison and Noel Gunther, for their careful, constructive criticism and their enthusiasm for tearing apart chapters that we had so painstakingly created; Marianne Harsanyi, for countless insights into the medical implications of genetic prophecy; Judy Williams, for her exhaustive comments on the drafts; the Rawsons, Eleanor and Ken, for their support and editorial direction; and Hugh Thomas, for the line drawings that accompanied the text.

We would also like to thank those whom we interviewed at length, whose points of view helped shape and embellish many of our thoughts and concepts: Thomas Bouchard, Barton Childs, David Comings, Leonard Heston, Phyllis Klass, Frank

Lilly, Daniel Nebert, Marcello Siniscalco, Michael Swift, and Judith Widmann.

Finally, our thanks to Susan Clymer, Liz Galloway, Irene Panke, and Beverly Smith, who worked long and hard to type and proofread the manuscript.

CONTENTS

PROLOGUE

This particular medical clinic seems much like any other, although it was obviously built by an institution in no financial trouble. The floors are covered with carpet, no linoleum. The walls are cream-colored instead of faded institutional green. The chairs are solidly framed in wood and cushioned, rather than the usual red-green-blue-yellow molded plastic, bolted to the floor. And the receptionist doesn't look as if she would kill you if you asked one question too many.

But the finer furnishings tell you nothing about what is actually going on here. For this is not just another medical clinic for allergies or orthopedic problems or heart patients. This is a waiting room for people who are about to undergo genetic screening—a first step into the future of medicine.

The two genetic counselors move around the room easily, spending time with those waiting, explaining the procedures. This particular clinic happens to specialize in prenatal screening—testing the fluid surrounding the unborn fetus for signs of genetic abnormalities like Down's syndrome (commonly called mongolism)—and for Tay-Sachs, a fatal genetic disease that attacks the nervous systems of about one in 3,600 Ashkenazy Jews and kills them before they reach the age of four. But prenatal screening is only one tiny apsect of the ability of the genes to foretell the future. The hospital complex also houses other laboratories with other tests, to help people make decisions on the kinds of work they can safely do, the ways they might live, the foods they might eat: all on the basis of what the genes reveal. This clinic is just one small jewel carved out of the enormous potential that is genetic prophecy.

As the morning goes by, startling stories filter out of the screening sessions:

- Upstairs in the hospital are two babies with Down's syndrome, both born to mothers who refused amniocentesis. One couple, in their forties, is cheerful and determined, happy to have produced a child—any child—to love and raise. The other couple, much younger, was advised to skip the prenatal test because the prospective mother was not in the high-risk group of women over 35. Their little girl is their second child. They are overcome with the impossibility of the situation, and with guilt.

- A pregnant woman who is to be tested for Tay-Sachs does not show up for her appointment. The counselors have learned that her husband has suffered a brain aneurysm and is dying. They wonder whether they will be able to test him for the deadly trait before he dies.

- A woman phones in, hysterical. She has a neurologic disease called neurofibromatosis which, in its most severe form, can cause paralysis. Over the past decade, a half dozen physicians have told her that the condition is not hereditary. Yesterday, another doctor informed her that the others were wrong. Neurofibromatosis *is* a genetic problem; the odds of her passing it on to her children are 50 percent—one chance in two; and she should think twice before she has any more children. She already has a nine-year-old son. She wants to know: Is he in any danger?

- The counselors discuss a couple in which the husband is a hemophiliac. Because any daughter he sires must be a carrier of the condition, the couple has decided to have children by artificial insemination. But they have also resolved not to tell their children of their unknown heritage, mainly because of family pressures. One of their two children is a girl, 18 months old. When she is ready to have a family of her own, she will undoubtedly undergo genetic screening, knowing of her father's condition. She will learn that she does not carry the trait for hemophilia, and therefore that she cannot possibly be her father's daughter. The parents do not yet understand the problem. They still view their child as a baby who must be protected, and not as the template of a mature young woman.

Those who come to this clinic undergo the same basic process as they would in any genetic screening. First, they provide a sample that can be tested. In this case, it is blood or amniotic fluid; other tests in other clinics might require a urine sample or a tiny swatch of skin. Then they go home and the laboratory begins its work, subjecting the sample to dozens of tests to discover the kinds of products its genes are producing, and, by inference, the kinds of genes the sample contains. When the patients return to the clinic a counselor will sit with them and explain the test results. For these prospective parents, the tests can reveal a great deal about the genetic health of their unborn children. In other clinics, workers may find out whether they are vulnerable to certain chemicals in their environment, patients may discover that they are susceptible to specific drugs, and the merely curious may learn that they carry a whole host of genetic traits that bear directly on the way they live. The counselors themselves are crucial to this process. They are the first contact that most people have with what may be the most profoundly personal of all the advances of the biological revolution. They are the translators of the vital secrets of the genes.

The clinic closes before lunch and the patients disappear. After the counseling sessions, it is up to them to decide how they will treat what they have learned. They can act on their new knowledge or ignore it. They can avoid things that may be harmful or accept the risks. Either way, they have discovered things that may influence the way they, or their children, choose to live.

PART 1
PROPHECY

PART I
PROPHECY

1

GENES AS PROPHETS

Declare the Past, Diagnose the Present, Foretell the Future.

HIPPOCRATES

It is the first thing asked, no matter who the person or what the problem. Man or woman, flu or cancer, arthritis or heart disease, the question is nearly always the same.

"Why me?"

Why, indeed, you?

The question is not melodramatic or rhetorical. It is not simply a statement of despair. "Why me?" is the question that medicine has been trying to answer ever since scientists learned that microorganisms could cause disease, since Sir Percival Potts first observed that only some chimney sweeps in eighteenth century England contracted scrotal cancer as a result of their profession.

"Why me?"—why one person becomes ill while another, also exposed to a disease, does not—is, in fact, the central question confronting medicine today.

The question is important primarily because medical science has done so well at finding out why people *in general* come down with a particular disease; literally hundreds of illnesses, from malaria to pneumonia, have offered up their secrets during the past century. It is central because the diseases that are rampant today—the cancers and emphysemas and heart problems—seem to strike almost at random, with medical understanding of the reasons often limited to "it runs in the family." It is critical because *preventing* illness, rather than curing it after it strikes, is becoming the watchword of modern medicine. "Why me?" means so much because if we can determine why one particular individual comes down with an illness, we might also learn to cut the chain of events that causes it, to avoid potential trouble spots in the environment, to know, exactly, who needs what kind of medical advice.

Until very recently, the answers medicine has offered us have not been entirely satisfactory. When a physician remarks that "You caught a bug," or "You work too hard," he is isolating the external factor—the germ or type of stress—that might make *anyone* susceptible. He is, in fact, not talking about you at all, but about how you fit into the statistics that have grown up around your problem. When a laboratory test discovers bacteria flourishing in your bloodstream, it is not pinpointing the underlying factors that may have allowed the colony to grow in the first place, but merely the existence of the disease.

Now, for the first time, the question "Why me?" can often be answered. During the past 25 years, we have learned that the secret as to why one person becomes ill while another, subjected to the same environment, stays healthy is partly contained in each individual's internal blueprint, the genes.

Genes never act alone. They are always influenced by the environment; they never confer absolute resistance to one disease or give an absolute guarantee that another will strike. In every single illness, the equation is the same: Disease occurs when an environmental insult meets genetic predisposition, when environmental and genetic factors come together.

It stands to reason, then, that if we can uncover both sets of factors, we can finally answer the question, "Why me?"

It also stands to reason that if we can test for the presence of genetic factors *before a disease strikes*, we can pinpoint who is at risk, foresee the probability that an illness will occur, and—by warning those who are susceptible to stay away from specific environmental triggers—actually prevent it from taking hold. And even when we cannot fully protect ourselves from factors that trigger some diseases—even when we are susceptible to such ubiquitous elements as automobile exhaust—we can still gain the advantage of an early warning. The genes can alert us to the probability that a disease is imminent; and forewarned is forearmed. It is an axiom of the medical profession that an early diagnosis often leads to more effective treatment and better odds for cure. Our newly found ability to read the secrets contained in the genes is one of the critical elements in making that happen.

The Rise of Prophecy

The Italian broad, or fava, bean is enormous; its pod can grow up to eight inches long. It sprouts with great enthusiasm on the borders of the Mediterranean, cultivated by farmers in Greece and Italy as well as by those on the countless islands that dot the sea. In spring and summer, it is a staple food for most of the local populations. Its seeds are dried and saved so that they can be eaten in winter.

For many, the fava bean is a tasty, nutritious part of the diet. But even in pre-Christian times the bean had its detractors. The Greek philosopher and religious reformer Pythagoras, for instance, barred his followers from eating it, or even from walking through fields in which it was growing. He kept the reasons for this prohibition to himself.

By 1900, people had begun to understand why the fava bean had been getting such bad press over the years. While the bean was perfectly fine for some to eat, for others it was turning out to be as deadly as a stick of dynamite.

For centuries, schoolteachers on the Mediterranean island of Sardinia have witnessed a curious phenomenon. Every February as spring arrives, some of their students suddenly seem drained of energy. For the next three months, their schoolwork suffers. They complain of dizziness and nausea and fall asleep at their desks. Then, just as suddenly, they return to normal and remain healthy and active until the next February rolls around.

In some countries, such incidents would be ascribed to boredom, spring fever, or a massive, collaborative effort to disrupt the learning process. But Sardinian adults suffer from similar symptoms; and while some merely feel a strange lethargy, others die after urinating quantities of blood. At times, as many as 35 percent of the islanders have suffered from this phenomenon.

In the early 1950s, scientists with no special interest in the Sardinians' particular problem came to the island to study its people. For them, Sardinia presented a special opportunity. Its isolated location to the west of Italy, its poor resources, and its rocky, inhospitable interior had long protected its essential character. Invaders such as the Phoenicians, Greeks, and Romans had occupied its low-lying areas; but when they departed, as they invariably did, the Sardinians returned to their old ways. For scores of generations, Sardinians have almost always married other Sardinians; even those who now take jobs in Germany and Italy come home to their native villages to marry. As a result, the island's people have evolved a relatively pure gene pool, protected by geography and uniquely influenced by the environment. Genetically, the island is like an oil painting that, despite minor restorations, retains its original colors and character.

For researchers like Marcello Siniscalco of the Sloan-Kettering Institute in New York, Sardinia provided a naturally controlled population—a living laboratory for human genetics. Siniscalco and others began to flock to the island to trace the hereditary patterns of disease.

At the same time, various scientific institutions outside of Sardinia were examining the origins of an odd disease called hemolytic anemia. One form of hemolytic anemia is hereditary in nature. It arises when the red blood cells actually begin to explode in the blood vessels. When the ruptured cells reach the kidneys, they are filtered out and excreted, causing the victims to urinate blood. If the amount of destruction is minimal, the loss of blood results in lethargy; if it is severe, the disease can kill.

Hemolytic anemia can have many origins. But in 1956, a scientific group from Chicago reported that almost everyone with the hereditary form of the disease was lacking a single enzyme called glucose-6-phosphate dehydrogenase (or G-6-PD), which forms a crucial link in the chain of energy production for the red blood cells. When the chain is cut, and when the cells need G-6-PD to defend themselves against certain chemicals, the cells can become more fragile; ultimately, internal pressure tears apart the weakened cell walls.

Clearly, the disease that so afflicted the Sardinians was hemolytic anemia. But the islanders became ill only in the spring, indicating that the victims' lack of G-6-PD was not setting the disease off by itself; something in the environment had to be taking advantage of the deficiency. The genetic factor may have been the loaded gun; but an environmental element was pulling the trigger.

Among the plants that flower during the Sardinian springtime is the fava bean. In the 1950s, the reasons behind its poor reputation became clear. Only those who both carried the defective gene for G-6-PD and ate raw or partly cooked fava beans (or breathed the pollen from a flowering plant) came down with the disease. Everyone else was resistant. Discovering the relationship of the enzyme, the beans, and the disease solved a problem that had been puzzling historians for centuries: namely, why Pythagoras forbade his followers to eat or go near the bean. It seems that Pythagoras himself was probably susceptible. (Pythagoras' aversion to the bean finally killed him. One day, a mob incensed at his religious beliefs supposedly found him at the home of one of his followers and began to chase him. Pythagoras ran, but as he was getting away, he reached the edge of a bean field. Consistent to the end, he refused to cross it. When the mob caught up with him, they cut his throat.)

Less than two years after the discovery of the links between hemolytic anemia, G-6-PD deficiency, and the fava bean, Arno Motulsky of the University of Washington developed a simple blood test to measure the presence or absence of G-6-PD. Armed with the test, scientists now had a way of determining exactly who was predisposed to the disease and who was not: The missing enzyme had become a tool of prediction; a signal that the disease might someday strike.

Motulsky sought out Siniscalco, and the two began to screen the Sardinian schoolchildren. Day after day they would enter a school, draw blood from a hundred fingers, and evaluate the samples in their laboratory, in school clinics, in hotel bathrooms when necessary. Gradually those in danger were located. They were warned to avoid contact with fava beans during the flowering season. As a result, the incidence of hemolytic anemia—and lackadaisical schoolchildren—began to decline.

Since that time, Motulsky's test has been refined and the impact of the enzyme deficiency has been examined more closely. Today it is known that about 100 million people around the world, including three million Americans, have G-6-PD deficiency, and that hemolytic anemia can be triggered not only by fava bean pollen but by a whole spectrum of other compounds, from antimalarial and sulfa drugs to aspirin and vitamin K. Because of a genetic marker—a gene that makes it possible to prophesy the future—many of those who are most susceptible to the disease can now consciously avoid the compounds that might cause them harm.

Crude forms of prophecy have always been a part of medicine. The Greeks could look at a frail or mongoloid child, accept the inevitability of its hopeless future, and throw it from a cliff; the German dye industry in the 1890s worked out the relationship between certain chemicals and the high incidence of bladder cancer among some of its employees; and modern physicians have long known that if a parent comes down with, for instance, diabetes, chances are greater that his or her children will come down with it too.

But the new genetic prophecy is not simply a modest improvement on this kind of general prediction. Instead, it makes use of genetic markers—the direct products of genes—to forecast the likelihood that specific diseases will occur. The realization that Sardinians were deficient in G-6-PD was used to predict their reaction to fava beans and was one of the first times a genetic marker was employed in that way. Now markers exist that can predict the possible onset of other diseases as well.

Markers already discovered have been linked to a fistful of diseases. Red hair, especially among the Irish, has been tied to high rates of skin cancer. A woman with blood type A who is taking oral contraceptives is five times more susceptible to blood clots than women with other blood types. People with higher levels of the digestive enzyme pepsinogen I in their blood are five times as likely as others to develop peptic ulcers. Smokers who are missing the protein alpha-1-antitrypsin are likely to develop emphysema nearly a decade earlier than nonsmokers missing the same protein. Still other genetic markers have been linked to everything from diabetes and arthritis to heart disease, malaria, and flu, while an entirely separate group seems to be tied to mental illnesses like manic depression and schizophrenia. Medical science is now on the verge of developing a comprehensive system of predicting and preventing diseases by analyzing each individual's own particular set of genetic markers.

The changes that genetic prophecy can offer are awesome. If a disease can be predicted before it strikes, the fetus conceived by a couple with a high propensity for multiple sclerosis can be tested for the relevant genetic marker—and may be aborted before the child, the family, and society suffer the inevitable hardship that would occur. Because the environment is so important to the onset of some diseases, those who live in polluted cities and are most susceptible to respiratory ailments or cancer can be located and forewarned. And because some markers may even help to predict the course a disease will take after it strikes, people who do become ill can better understand what is happening to them; they and their doctors can then act accordingly.

These possibilities already exist for some diseases. There are today over 200 genetic centers in the United States that screen for specific ailments. Some centers have already incorporated tests for a few genetic markers into their repertoires. Other tests are still in the experimental stages; but it will not be long—perhaps within the next 10 years—before any one of us can go to the local clinic, have a blood sample drawn, and receive a computer print-out of our susceptibilities to scores of diseases; before an expectant mother can provide a sample of her fetus' blood and learn not only whether the child will contract any of the 60 or 70 genetic diseases now detectable, but how, later, she might raise it in the healthiest possible environment; before an industrial concern can determine which of its workers may be in danger and provide them with safer jobs or clean up the work environment to protect them. Gradually, the genes are becoming the focal point for our understanding of what disease is and how it works. More and more, we are beginning to rely on the stories they tell to decide how we can best establish and maintain good health.

The Gene Hunters

In the middle of the nineteenth century, a German cytologist named Walther Flemming took a sample of cells from salamander testes, dyed them, and placed them under his microscope. As he squinted through the eyepiece, he noticed for the first time that clots of material in the cells' nuclei had absorbed the dye especially well; they were starkly visible against the otherwise colorless background of the cells' bodies. Some of the cells in his samples were in the process of reproducing and dividing. In them, Flemming found that the colored material had separated into distinct, slender, thread-like strands. Soon other scientists noted the same phenomenon. One of them, W. Waldeyer, dubbed the stained material "chromosomes"—colored bodies.

That simple experiment unveiled one of the most important structures of genetics and heredity; for, as we now know, chromosomes are nothing more than long chains of genes attached end to end. The genes themselves contain the record of the body's past, as well as a blueprint and map of its future.

Chromosomes are composed of long, twisted chemical threads of deoxyribonucleic acid, or DNA. In cells, the twisting pattern of the genes is very regular. DNA, appearing much like an endlessly spiraling staircase, forms one of the most glorified structures in science today: the double helix, a simple, elegant coil that is the basis for all life on earth. The double helix itself is a bit inscrutable. But if we could remove a strand from a cell, dry it out, and untwist it, the molecule would look like a simple ladder, with two long sidepieces supporting a regular series of rungs. The sidepieces are standard, uninteresting. But the rungs are different. They come in four different types—each containing a different signal, like the dots and dashes in a molecular morse code. It is by reading the message carried by the rungs that the cell knows what it is to be and how it should act.

The information in the rungs is translated by a complicated process into proteins, the tiny molecular building blocks that are foundations of both structure and function for the cells. Each gene—each complete coded message—produces a single protein. And the proteins assemble to take part in the creation of blue eyes, bones, nerves, organs, and muscles, as well as in the development of the massive communications network that keeps all the parts of the body running smoothly and assigned to their proper places. In all, each cell of the body contains about 100,000 individual genes.

But the DNA that carries these genes is coiled, twisted, doubled back upon itself, a complex mass of blueprint. If it were stretched out to its full length, it would be just over 6½ feet long. The 100,000 functioning genes manage to take up only a tiny portion of the genetic material available. A portion of what is left consists of copies of these genes, the same basic code, repeated again and again with minor alterations, so that the cell can produce the same enzymes and proteins under different sets of environmental conditions. The result of this repetition and overlap is a fantastic degree of genetic variation, all of which allows the cell to adapt to a vast array of possible challenges.

Each cell contains all the genetic information that is needed to build and monitor every genetic structure and function in the body. But the requirements of, say, a nerve cell and a cell that produces insulin in the pancreas are completely different. For that reason, the rest of the genetic material in the chromosomes is dedicated to regulation, to acting as genetic stop-and-start signals that determine exactly when a gene is going to produce a protein and when it will be shut down. The amount of regulatory material needed to oversee and direct the function of every cell in the body dwarfs the amount needed for the genes themselves; scientists have estimated that the genetic blueprint for regulation is about one hundred times as extensive as the map of the genes.

Together, the genes and regulatory material function like some complex, remote city of old. The scientist seeking to learn the nature of its operation becomes a kind of genetic archeologist, chipping away at the unknown. Gregor Mendel, the obscure Moravian monk who first described the gene in 1865, postulated the city's existence. Francis Crick and James Watson, who proposed a structure for DNA in 1953, sketched the first rudimentary map. Now scientists have invaded the city proper, unraveling the codes of single genes, finding out what protein each one produces, exploring where each gene sits on the chromosomes.

At the same time, they are finding clues as to how the genetic city operates. What they are discovering is a monumental paradox. The basic genetic laws are simple, but when genes and proteins interact, they do so not in a straightforward, practical arrangement, but through a spaghetti-like tangle of conflicting signals, each with its own purpose, each linked to others in ways that become clear only through painstaking investigation. As the scientists explore, they are discovering what role each gene plays in life and how, in the long, complicated string of operations needed to translate a gene's message into a product, things can go wrong. At the same time, they are beginning to decipher the signs and signals that tie the city together. In the process, they are uncovering genetic markers.

The theory behind genetic markers is simple: All diseases

have genetic components; if we can learn what they are and can test for them, we can find out who is susceptible to what disease.

In practice, however, the process is more difficult. Someday, genetic engineering will allow us to identify each of our genes directly; in fact, researchers can already isolate the gene for sickle-cell anemia. But we are still a long way from being able to use that kind of clinical analysis of all our genes on a regular basis.

If genes cannot yet be easily identified, however, their products can. And that is the secret of genetic prophecy. A genetic marker is the recognizable characteristic—for example, the presence or absence of a protein—by which we can detect the existence of a gene. The marker can be the direct cause of susceptibility to disease, as the deficiency of the enzyme G-6-PD is for hemolytic anemia. But it can also be a signpost, so closely linked to the actual cause of a disease

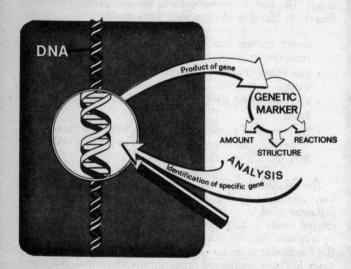

1. Genetic markers—the key to prophecy. Each functioning gene helps produce a single distinctive protein. Scientists can use these proteins as markers, identifying them, analyzing their characteristics, and discovering how they influence our responses to environments. Ultimately, they can identify the original gene, simply by looking at its product.

that it almost invariably indicates the presence of some other as yet undiscovered marker. In the 1950s, for example, the same Sardinians who became ill after eating fava beans were invariably color blind. Yet the genetic characteristic of color blindness has nothing to do with hemolytic anemia. As it turns out, the gene that results in color blindness in Sardinians sits right next to the gene that causes a deficiency of G-6-PD; they are almost always inherited together, generation after generation. Even if the genetic basis for hemolytic anemia had never been discovered, therefore, scientists could have used color blindness as a reliable genetic marker to pinpoint those in the Sardinian population who were at risk.

By now, hundreds of markers have been identified for hundreds of diseases. As a result, the medical profession, which, for centuries, has been treating many diseases as purely genetic or purely environmental, is slowly revising its approach. The old way of viewing diseases as separate entities is out. The new way—recognizing the complicated nature of illness—is taking hold. The basic facts are these:

- Every disease has both environmental and genetic components.
- For every genetic component, one or more markers exist.
- Environmental factors can be tracked down and isolated.
- No portrait of a disease is complete without a thorough understanding of both genetic and environmental factors and the way they influence its course.

The relative weights of environmental and genetic factors vary according to the disease. Phenylketonuria (PKU), for instance, used to be considered a purely genetic disease that caused mental retardation. Now we know it to be controllable by a special diet, an indication of the influence and impact of the environment on the disease. Heart disease, on the other hand, has been considered primarily environmental. But recent studies have found that, while everyone is susceptible to heart disease in some way, some are more susceptible than others. Those who are *resistant* require a large dose of environmental danger before they become ill—by eating the wrong foods, perhaps, refusing to exercise, and leading a particularly stressful life. Those who are *susceptible* may become ill even if they are careful; they are predisposed to the disease as long as they receive minimal environmental encouragement. For everyone, the right genetic marker can point to the *probability* that a disease will strike. The only real limits to prophecy are the number of genetic markers a particular disease might have (the more it has, the more complicated prophecy becomes) and the accuracy of the tests that uncover them.

The Longevity Syndrome

Recently, Charles Glueck of the University of Cincinnati College of Medicine reported an astonishing discovery. He had managed to uncover two groups of people who seem to be genetically endowed with longer or shorter life spans than the general population. One group lives an average of 5 to 10 years longer than average. They rarely contract atherosclerosis, or coronary heart disease, no matter what they eat or drink. The other group suffers from the opposite set of circumstances. They carry a much higher risk of heart disease than average; and they live shorter lives.

Heart and blood ailments are the greatest killers in the United States today. They include high blood pressure (or hypertension), coronary heart disease, rheumatic heart disease and strokes. In 1980 alone, they accounted for nearly one million deaths—over twice the number attributed to cancer, accidents, pneumonia, influenza, diabetes, and all other causes combined. And they affect another 39 million Americans who are still alive. Although a growing awareness of the environmental factors involved in heart disease has begun to lessen its impact, about a million and a half Americans will suffer heart attacks this year.

All this illness is not cheap. The total estimated cost of cardiovascular disease in the United States for 1981 amounts to over $46 billion for physician and nursing services, hospital care, medication, and working time lost because of disabilities. The American Heart Association alone has thrown over a third of a billion dollars into research in the last 30 years.

Until recently, heart disease was considered an "environmental" disease—that is, seemed to be explained largely by the fact that those who became ill generally smoked, ate diets high in cholesterol, refused to exercise, had high blood pressure, or lived under the weight of a combination of these factors. But now the genetic elements of heart disease are beginning to surface. As research fine-tunes our understanding of the problems, the *complementary* nature of their genetic and environmental components is becoming clear. Physicians are beginning to realize that smoking, diet, and a lack of exercise may be risk factors, but risk factors for whom? The genes seem to make some people especially vulnerable.

Perhaps the most interesting piece of news discovered so far concerns that old villain, cholesterol. Although all the scientific evidence is not in, most scientists now agree that cholesterol by itself may have little impact on health or heart disease. While high levels of serum cholesterol—or cholesterol in the bloodstream—may or may not signal a predisposition to heart disease, they are probably not the key to sound prediction. They are, instead, mere baggage for the real actors in the drama; they are capable of causing problems, but only under specific conditions.

It is now known that cholesterol does not simply rush through the bloodstream alone, clogging arteries along the way. It travels linked to several kinds of blood proteins. One kind, called high-density lipoproteins (HDL), actually *reduces* the risks of heart disease; another, called low-density lipoproteins (LDL), *increases* them.

To bring oxygen to all parts of the body, the heart operates its own private monitoring system, pumping about 100,000 times a day, pushing some 4,300 gallons of blood through the blood vessels. It also forces the cholesterol-protein combination (which, in moderate doses, provides food for cells) throughout the body. About 80 percent of the lipoproteins carrying cholesterol are LDL; most of the other 20 percent consists of HDL.

LDL tends to act like the body's dump truck. It carries cholesterol along the blood vessels and deposits it along the way, which is fine if the level of cholesterol is exactly what the cells need. But if LDL deposits too much—if its concentration in the blood is too high—the excess cholesterol accumulates. The accumulation narrows the blood vessels and triggers atherosclerosis and heart disease.

HDL, on the other hand, is the circulatory system's vulture. Although scientists are still not sure exactly how it works, it seems to act as a scavenger, grabbing cholesterol that is piling up and carting it off so that it can be excreted. HDL may also block the cells' uptake of LDL cholesterol; the cells can still receive what they need from HDL, but in a lower concentration.

HDL was first studied in 1951, when a group of men suffering from coronary heart disease were found to have low levels of the protein. But HDL was difficult to measure; it seemed like a minor component among the lipoproteins; and studies of the total cholesterol level and LDL looked as if they were beginning to pay off. As a result, these early findings were largely ignored.

Not until 20 years later did researchers again take a good long look at HDL. What they found provided powerful evidence for its role in protecting against heart disease. Quickly they confirmed the crude studies of 1951. Independent surveys of Japanese-Hawaiian men, black sharecroppers in Georgia, Israeli men, and the entire town of Framingham, Massachusetts, statistically showed that low concentrations of HDL led to higher risks of heart disease, independent of all other risk factors, including the levels of LDL. By now, further studies have made one conclusion inescapable: Just measuring total cholesterol levels is not enough to diagnose the risk of heart disease. Two people with the same levels of cholesterol may have completely different proportions of HDL in their blood. The percentage of HDL must be measured separately.

That HDL and LDL should account for the differences among the life spans of Charles Glueck's groups should come as no surprise. Glueck found that those who lived longer were genetically endowed with either high levels of HDL or low levels of LDL; he labeled these characteristics "the longevity syndrome." And those at higher risk of heart disease? They carried genes that left them with high levels of LDL or low levels of HDL.

Altogether, Glueck's groups comprise about 3 to 5 percent of the population. For the rest of us, the genetic contribution to our levels of HDL and LDL is far more subtle; the lipoproteins seem to be extremely sensitive to environmental factors. Exercise and moderate drinking (one to two highballs a day), for instance, have been statistically linked to higher levels of HDL; cigarette smoking and obesity are associated with significantly lower levels.

Already researchers are looking at the possibility that HDL and LDL levels should be screened nationwide. Robert Levy, the director of the National Heart, Lung, and Blood Institute in Bethesda, Maryland, has pointed out the value to some of having their HDL levels measured. Most people who have high levels of cholesterol also have high levels of LDL. But some have high cholesterol and high levels of HDL. Because high cholesterol has traditionally led physicians to recommend drugs and a low-fat diet, distinguishing between those who really need it and those who are not necessarily at risk is critical. People with high levels of HDL "could be spared a lot of grief," notes Dr. Levy. And those with higher levels of LDL could be given treatment designed to match their level of risk.

Tests for HDL levels do exist. But they are not yet available for mass screening. Most physicians hope, however, that within five years we will be able to pinpoint with far greater accuracy those whose levels of cholesterol put them at greater risk of coronary heart disease. They also envision the possibility of controlling an individual's level of HDL—of raising it if necessary—with programs that are far simpler and less unpleasant than the diets and drugs that are now used to reduce general cholesterol levels. One note of warning, however: while moderate drinking seems to raise HDL levels, it also has its share of unpleasant side effects. Two drinks a day to reduce the odds of coronary heart disease can result in damage to the liver and other forms of heart disease.

The most common ailment of the circulatory system is not coronary heart disease, but hypertension (high blood pressure). Hypertension is also one of the most telling prognosticators of impending coronary heart disease; it is, in fact, a marker in itself. And researchers have discovered that it may have an important genetic component.

In 1974, scientists at the Brookhaven National Laboratory in Upton, New York, succeeded in breeding two groups of rats, one susceptible, the other resistant to hypertension. Then they subjected both groups to psychological stress. The rats with genetic susceptibility showed persistent elevations in their blood pressure; the rats that were resistant did not. The conclusion was that hypertension has a genetic component, and that it can be brought on by psychic stress.

Although no accepted marker yet exists for predisposition to hypertension, the disease itself, as a marker for heart and kidney failures and stroke, can be diagnosed. Furthermore, effective forms of treatment do exist; hypertension is now relatively easy to bring under control.

Medical Futures

Genetic prophecy is best played as a game of averages. The object is to consistently narrow the focus of attention, to use markers that increase the accuracy of prediction as they pinpoint those who are the most susceptible. In the case of G-6-PD deficiency, for instance, the first, rough estimate of those who may contract hemolytic anemia takes in the entire population of Sardinia. From there, the focus begins to narrow. The second level of prediction is restricted to Sardinians who come in contact with fava beans; the third is even more defined, limited to Sardinians who eat the beans (or inhale their pollen) and are saddled with the specific enzyme deficiency. The art of predicting who may be susceptible to heart disease in the United States works the same way. The largest group at risk is the American population; since over 40 million people suffer from heart disease at any one time (out of a population of nearly 230 million), any one individual has a 20 percent chance of becoming ill. From that point, the group becomes progressively smaller and the odds of illness increase. The risks are greater for smokers, for overweight smokers, for overweight smokers with high levels of LDL, until, when all the factors are accounted for, the group with the highest risk can be pinpointed. In the same way, that small segment of the population with the genetic propensity for high levels of HDL, is the focal point of *resistance* to heart disease.

In the end, research into genetic markers is striving for these major goals: to place multifaceted problems like heart disease in the same perspective as such straightforward genetic ailments as sickle-cell anemia; to increase the accuracy of prediction to the point where a marker offers reliable information if the right environmental factors exist; and to continue to narrow the size of the group that is the object of prediction until it involves a population of one, you.

Genetic prophecy is in the cards, not in the twenty-first century nor even in a decade or two, but now. Research is progressing at a breathtaking pace; markers found in the laboratory just months ago are already being introduced into clinical practice. Those now in use go far beyond G-6-PD; they apply to people in every environment, of every race, in every occupation.

Genetic markers, of course, can take us only as far as the inherent laws of genetics will allow. But the limits of those laws are being stretched daily. Stripped of all its trappings, the most important element of human genetics is its power of prediction. To the patient who asks, "Why me?" and the physician who wonders, "Will this disease strike again?" genetics offers answers. For genes *can* foretell the future; and it is only a matter of time and technical skill before we understand their language.

For medical treatment in general, the influence of genetic markers will continue to grow. Every time we uncover the genetic components of a particular disease, we take another step toward learning to control it. As such, genetic prophecy will help us to treat the *causes*, and not merely the symptoms, of diseases that have resisted treatment in the past. In the process, our systems of treatment will become more effective.

For individual patients, the prognosis is just as bright. By predicting the possibility that disease might occur under certain environmental conditions, genetic markers will make it possible to prevent many diseases that, today, seem untreatable. If, for instance, someone knows that he is genetically predisposed to fatty deposits in his arteries, he might well choose a life that avoids substances and stresses that increase the chances of heart disease. And even if he suffers a heart attack despite his precautions, the medical community will have been forewarned about his susceptibility; the problem can be diagnosed in its earliest stages; and the treament can begin when his odds for survival are the greatest. As such, genetic markers constitute a kind of early warning system, a network of signals that can foresee impending danger, offering the physician a chance to deal with a problem before it requires the drastic, destructive therapies that are sometimes needed in the latter stages of disease.

For society, genetic markers can play a variety of roles. Medical costs and insurance rates, so high because of the expense of extended hospital stays and the complex treatments generated by new technology, may be slashed. Workers may be protected from chemicals that endanger them.

The percentage of those who spend the last years of their lives as invalids—who, in essence, swell the ranks of the handicapped after radical forms of surgery and therapy—may be drastically cut. And fetuses that have a propensity for a particular disease may be treated even before they are born, while they are still in the safe, protected environment of the womb.

In all these ways, genetic prophecy will offer a new and better way of practicing medicine. As we learn how to use it, we will begin to reduce our overdependence on high medical technology, counting on prediction, prevention, and early diagnosis as our first line of defense. We will recognize a clearer concept of disease itself, as something waiting in our genes, ready for the right trigger to spring it free. And we will begin to work harder at staying healthy in the first place.

An old Chinese proverb states: "A poor doctor cures; a good doctor prevents." It is a proverb that medicine is taking to heart, and genetic prophecy is helping to lead the way.

2

THE ENEMY CHANGES

It is much more important to know
what kind of patient has a disease than to know
what kind of disease a patient has.

CALEB PERRY
18th Century Physician from Bath

The idea of preventing disease has been around for a long time. But its acceptance into general medical practice is a recent phenomenon.

Ever since the earliest beginnings of modern medicine in the halcyon days of the Greeks, two different and opposite concepts of disease have struggled for supremacy. The first, known as the Platonic view, treats disease as something that attacks healthy people more or less at random. As such, each disease has a name and a separate identity. We expect those who contract it to comform to a set of recognizable characteristics: pain in the chest, perhaps, a hacking cough, fever, runny nose, sore throat, burning eyes. Medical students are taught to look for specific groups of symptoms to diagnose a disease. But they learn that the list they memorize represents a "textbook case," and not necessarily the medical portrait of an actual patient. Real people, it turns out, usually manifest only parts of the list. In what is called a "mild common cold," for instance, the widely accepted *Textbook of Medicine* notes that 97 percent (not all) of patients actually suffer sneezing; 49 percent, feverishness; 43 percent, chills; 28 percent, burning eyes and mucous membranes; and only 22 percent, muscle aches. In other words, that supposedly well-defined entity, the cold, bears a striking stamp of individuality, depending on who becomes sick.

The second concept, known as the Hippocratic view, treats disease not as an external force but as a deviation from the norm. In fact, those who subscribe to it try to avoid the use of the term *disease*, preferring to refer only to sick people, to those who have had trouble adapting to particular conditions at a particular time. According to them, *certain* people under *certain* conditions are unable to cope with an environmental insult; their bodies simply cannot adjust. The result is illness.

A few early physicians fought hard for the Hippocratic concept of disease. In the tenth century, the Arabian physician Rhazes said of those disposed to smallpox and measles: "Vulnerable bodies are generally such as are moist, pale, and fleshy; the well coloured also, especially if they are ruddy and tending to brown, are disposed to it, if they are loaded with flesh. So are likewise those, which are frequently liable to acute and continual fevers, to running of the eyes, red pimples, and boils. . . ." He recognized that differences in people could *predispose* some to disease and determine its severity and character.

The Reign of Terror

Despite the observations of Rhazes and others, the Hippocratic concept had few adherents. Most found good reason to ignore it. Until 80 years ago, when we began to learn how to control infectious diseases, death nearly always seemed to strike from outside the body. From the days of the earliest cavemen, death was usually violent, caused by something foreign; and disease seemed just as alien. The nineteenth century painter Arnold Bocklin depicted *The Plague* as flying into a village on the back of a serpent. Others described death as an unwelcome and unexpected visitor, knocking on its victims' doors.

These concepts were not mere superstitions. They were not simply founded on ignorance. Most illnesses and deaths were, in fact, attributable to accidents or to agents of disease which could be spread from person to person. Life and health were constantly threatened by uncontrollable violence from without—by earthquakes and wars, by illnesses and accidents.

And if these dangers were not enough, mankind periodically faced devastating attacks from pestilential diseases that took on the entire world.

In the year 541 A.D., a murderous epidemic, known as the Plague of Justinian—the first wave of the Black Death—reared its ugly head. It took hold and ruled the civilized world for more than fifty years.

Procopius of Caesarea recorded what he saw then in Constantinople: "During these times there was pestilence, by which the whole human race came near to being annihilated . . . It started from the Aegyptians who dwell in Pelusium. Then it divided and moved in one direction towards Alexandria and the rest of Aegypt, and in the other it came to Palestine on the border of Aegypt; and from there it spread over the whole world, always moving forward and travelling at times favorable to it. For it seemed to move by fixed arrangement, and to tarry for a specified time in each country. . . ." The perception that disease was an external evil was reinforced by Procopius' observation that it was not uncommon for those about to be inflicted to see demons. The plague's future victims would lock themselves in their houses so that the agents of disease could not get at them. But such visions were actually early signs of illness; for them it was too late.

The plague reached Italy and France; by the end of the sixth century, half the population of the Byzantine Empire had perished. Procopius wrote that in Constantinople "the tale of the dead reached five thousand each day, and again it even came to ten thousand and still more than that." Somehow mankind survived.

In 1338, the Black Death reappeared in Central Asia. Within a decade it had moved west to France and England. On April 27, 1348, a cantor visiting France wrote to his friends in Belgium: "To put the matter shortly . . . more than a half of the people of Avignon are already dead. Within the walls of the city there are now more than 7,000 houses shut up; in there no one is living, and all who have inhabited them are departed; the suburbs contain hardly any people at all."

The plague was vicious and insidious. It killed without mercy, leaving so many dead that the streets were foul with the smell of rotting bodies. It destroyed whole populations until there was practically nobody left alive to spread the disease. Then it would disappear, only to surface somewhere else. Between 1350 and 1600, it materialized some thirty separate times in dozens of different places. And in 1655, it roared down upon London in a massacre that is known today as the Great Plague. The epidemic ravaged not only that city but the neighboring villages as well. London began to look like a ghost town as people fled. But no place was safe. The villagers of Eyam, a nearby community, saw their numbers dwindle from 350 to only 30—a death rate of over 90 percent. Eyam's preindustrial version of the Andromeda strain most likely arrived from London in a box of contaminated clothing, for the town's tailor, who had opened the package, was the first to die. As the villagers started to fall ill, some tried to flee, but they, too, died. The local rector, Reverend William Mompesson, wrote in despair: "As each family left the village seeking sanctuary from the plague, they would carry with them, hidden in their baggage, among their garments, upon their hands and lips, the invisible seeds of disease. Sickness and death would travel with them, as unseen companions, measuring their progress step by step. Wherever they wandered, through whatever town or hamlet they passed, in whatever house they sank to rest, the Black Death would follow, like a terrible shadow."

Plague was not the only horror then. Epidemics of cholera, smallpox, yellow fever, typhoid, diphtheria, malaria, and countless other diseases also took their toll. Some even competed for victims; diseases that appeared on the scene of some other killer's devastation faded quickly because so few people had been left alive. Together, the epidemics reaffirmed the perception that disease was the result of external forces that attacked and invaded helpless and hapless victims.

Controlling the Killers

The first scientific attempts to stem the tide took place in the eighteenth century. In the 1790s, William Jenner, an English physician, learned of a curious legend in folklore: milkmaids who had been exposed to fluids from cowpox ulcers rarely came down with the disease's human form—smallpox—even in the midst of severe epidemics.

Jenner found cases in which he could confirm the legend. Then, to test its validity, he took a healthy young boy named

James Phipps, scratched his arm with a needle, and rubbed in some pus taken from a milkmaid's sores. Two months later, he inoculated Phipps with the real thing—pus from a smallpox victim, enough to obliterate someone who was unprotected. The boy showed no adverse effects whatsoever.

Jenner's method of preventing smallpox spread quickly. Thomas Jefferson vaccinated his family; physicians in India tried to conduct wholesale vaccination of villages; and other scientists in England struggled to refine the technique to increase its effectiveness and safety.

Protection and prevention. The concept was brilliant, but far from new. Shamans and witch doctors had always used herbs and chants and the invocation of spirits to protect their clients from disease. But now someone had uncovered a chemical factor that could interact with the individual to make him immune. That single discovery laid the foundation for a scientific explanation. It pinpointed the importance of both the external agent (the disease) and the individual's physical state in the onset of illness.

Unfortunately, while people recognized the importance of vaccination for a single disease, it took nearly a hundred years before the *principle* of vaccination was used to combat other infectious diseases as well. Not until Louis Pasteur and Ignatius Semmelweis demonstrated how infectious agents could cause disease did the science of prevention progress.

At the turn of the twentieth century, tuberculosis, syphillis, typhoid fever, dysentery, whooping cough, diphtheria, influenza, and diarrhea together still accounted for 647 deaths for every 100,000 people each year. Improvements in sanitation, housing, nutrition, and medical care began to slash that rate. By 1920, the number was down to 430; by 1940 it reached 147; and by 1959, only 41 people out of 100,000 succumbed to those diseases. According to the Surgeon General's Report on Health Promotion and Disease Prevention, if conditions today were equal to those existing in 1900, we would have 400,000 deaths in the United States from tuberculosis, 300,000 from gastroenteritis, 80,000 from diphtheria, and 55,000 from polio. Instead, the deaths from *all four combined* total only 10,000.

Today, the once-great infectious killers are themselves dying. On September 3, 1979, the World Health Organization proclaimed smallpox to have been eradicated. Thanks to both preventive and interventive methods, most infectious diseases are now under control. It is unlikely that another scourge such as the influenza pandemic of 1918, which claimed 20 million lives, will ever sweep the world again.

To be sure, acts of prevention, such as smallpox vaccination and the killing of malarial mosquitoes, emphasized the importance of the interaction between an agent of disease and the individual. But our success in fighting infections kept our attention focused on external "causes." Whether vaccination prevented a disease or antibiotic therapy cured it, the target was still viewed as something external that could be isolated, examined, characterized, and destroyed.

Because of its apparent simplicity, microbial infection has remained a popular model for all disease. But that simplicity is also the very reason why the Platonic view of disease fails as a model today. As Barton Childs, Professor of Pediatrics at the Johns Hopkins School of Medicine, has pointed out, in the Platonic view, "differences between cases are perceived as variations on a single theme. But all our experience tells us that in fact diseases are not unitary in either cause or expression." In other words, the fact that the Platonic view is simple and easy to understand does not make it right.

Unfortunately, the Platonic view has been difficult to shake. It has held on despite evidence confirming the mutual importance of genes and environmental factors in disease. It has remained a part of our medical philosophy even as the great plagues of the past have been conquered, and is even now reflected in the language we use to describe our most important efforts to fight disease, our "war on cancer," our "battle against heart disease." And it is reinforced by our continuing desire to lay the blame elsewhere, to set up a relationship between ourselves and disease that portrays us as the heroes in the white hats, as fighters pitted against forces that can and will be overcome, given the right combination of time, money, luck, skill, and effort.

But that attitude, which worked for diseases in the past, is no longer good enough. Our enemy is changing. Now that the proportion of acute, infectious diseases has dropped, our most important medical problems are chronic, debilitating diseases—heart disease, cancer, arthritis, diabetes—which, together, cause 75 percent of all deaths in the United States. To fight these diseases, medical professionals are being forced to revise their way of thinking. They have had to begin to discard their obsession with building a better mousetrap for disease; now they are beginning to try to create a more congenial mouse.

Physicians are slowly arriving at the conclusion and are beginning to accept the Hippocratic view of disease, which is flexible and complex. It emphasizes individuality and the existence of multiple causes for an illness. And it demands that medicine become aware of how closely those causes are linked to each person's biological individuality and way of life.

In the Hippocratic concept, the causes of illness are not harmful unless they are combined with an individual's susceptibility. The swine flu virus can exist happily for generations in unaffected herds of swine. But if it is transmitted to humans, it triggers a dangerous illness. And the famous Typhoid Mary spread the bacterial agent of typhoid fever to countless others while remaining symptom-free herself; her defenses could accommodate and adapt to any changes the bacterium tried to produce.

The Hippocratic view has been buttressed by our growing recognition of the impacts that so-called genetic diseases have on our lives. Consider these facts:

- According to the Department of Health and Human Services, over 15 million Americans suffer from one or more types of birth defects, 80 percent of which are thought to be caused by genetic changes.
- Fifty percent of all miscarriages and at least 40 percent of all infant deaths are attributed to genetic factors.
- As many as 30 percent of all pediatric and 10 percent of all adult hospital admissions stem directly from genetic disorders.
- Nearly 3,000 genetic diseases have already been identified and catalogued.
- The life-years lost to these diseases are estimated to be six and a half times as many as those lost to heart disease.

Clearly, genetic diseases are, collectively, one of the biggest health headaches in this country today. But they merely illuminate the more obvious effects of genes. Other, more subtle effects crop up in every other disease as well. In fact, from the Hippocratic point of view, health might best be described as a balance in the delicate interplay between the genes and the environment. A healthy body stays healthy if it can respond and adjust to changes in the environment; a body becomes ill when environmental insults outstrip its ability to react.

The Platonic disease model worked well enough for infectious diseases, mainly because they are triggered by such obvious environmental influences. But it cannot explain why some people come down with chronic diseases while others do not. And it is particularly useless in helping us understand why chronic diseases occur; for they do not always involve obvious external agents that are analogous to the viruses and bacteria of infections. Decades of research have gone into isolating the "causes" of cancer; viruses, radiation, and chemicals have all been singled out as primary suspects. But isolating a single cause has proved impossible; cancer seldom strikes unless a combination of factors—both internal and external—are present. Those who try to explain it otherwise resemble the three blind men who, in trying to picture an elephant after touching its various parts, each comes up with a different description: the one who feels the trunk describes the animal as long and flexible; the one who feels the leg assumes that it is tubular and sturdy; the one offered the tail believes it to be thin and pointed.

The medical establishment's insistence upon holding to the Platonic point of view has slowed our progress toward arresting chronic diseases. After decades of research on diabetes and hundreds of millions of dollars spent to isolate its causes, a sugar test of the urine remains the best diagnostic test, the daily injection of insulin, the best treatment for more severe cases. After a war on cancer that has spanned the terms of five Presidents, our success in conquering most of its forms is still minimal. As a result, the public has become disenchanted. The money it has poured into research seems to have achieved few positive results.

This dissatisfaction is now compounded by the soaring costs of health care. Since 1960, the medical profession has managed to increase the life spans of Americans by about three years. A significant proportion of that number is due to a 25 percent reduction in infant mortality. In the same period, the nation's total spending for health care has rocketed from $27 billion to over $200 billion annually, an increase of about 800 percent, and its proportion of the nation's gross national product has grown from about 5 percent to more than 9 percent—or nearly $900 for every man, woman, and child in the country. Today, health expenditures take 11 cents out of every federal dollar—money that is generally funneled into attempts to treat disease and disability, rather than to prevent them. Since chronic diseases tend to respond poorly to treatment—and must be prevented to be "cured"—the medical services offered to those who suffer from them are often purely palliative, geared toward reducing pain, shoring up the body's defenses, delaying death, and, at times, prolonging personal agony. While these efforts may be worthy of praise, they do not contain the answers. In fact, they hardly touch on the crucial questions.

Sharing the Responsibility

One critical issue facing medicine today is the question of whether its role in our health and well-being should be drastically changed. Many theories exist as to why we live as long as we do, and why we might not be able to live much longer, no matter what the state of medical science. Until recently, most researchers believed that advances in medicine would continue to increase our life spans, accelerating the trend toward an older, weaker population with a greater dependence on expensive health care. But now scientists are beginning to believe that each of our cells contains a kind of biological clock, a genetic timepiece that, after a set number of years, simply runs down. The clock is thought to be controlled by any number of factors: by mistakes in the translation of genetic information which saddle the cell with defective proteins; by the apparent ability of cells to divide only a limited number of times before they die; by chemicals produced in the body that seem to trigger the signs and symptoms of aging. Most gerontologists subscribe to several of these theories, believing that various factors all contribute to aging and prevent us from becoming immortal.

Someday, perhaps, we may learn to rewind the clock (or at least to slow its ticking) so that a measure of immortality lies within our grasp. Until then, we are locked in to a limited lifespan, one past which few people live. The hypothetical upper limit for the *average* person is thought to be between 85 and 95 years old. And although there are those who live beyond that (the longest recorded and verified life span is 114, in Japan), that is the age to which most of us can realistically aspire.

There are already groups among our population who come close to what seems to be the ideal in longevity. Caucasian women in the United States and other industrialized Western civilizations have average life spans, at birth, of over 76 years of age, which is within a decade of the lower edge of the biological limit. And while other groups—white men and racial minorities of both sexes—have lower averages, we are all much closer to the theoretical limit than we were, say, in 1900, when the average life span was about 47.

As a result, while modern medicine can and will continue to extend our life spans, its primary objective is slowly changing. If longevity is, indeed, limited, medicine's objective must be to make those later years count, to offer us a fuller, healthier existence as we grow older. Instead of simply giving us quantity, it must deliver quality; instead of adding years to our lives, it must concentrate on adding life to our years.

This change is not just a matter of the emergence of genetic prophecy, although the tools of prophecy certainly help by allowing the medical profession to act *before* the onset of illness, rather than by confining it to intervening in disease only after it has already set in. It is, rather, due to the increased emphasis on preventive medicine, the strengthening of the belief that it is simpler, cheaper, and more effective to stop disease from occurring than to combat it once it breaks out. Originating in a small medical circle, the move toward prevention has been buttressed by a general movement toward medical responsibility. "More prople are taking responsibility for improving their health," says Gretchen Kolsrud of the U.S. Congress' Office of Technology Assessment. "Physical exercise is on the rise and people are learning how to manage such potentially dangerous factors as stress in their lives. We seem to be trusting less in others or in fate and looking more to ourselves."

The apparent inability of medicine to respond to the change is collaborating with this new awareness to help begin what may, in the future, amount to a health revolution. The public's impatience is making itself felt. Physicians are being pulled from their pedestals. Their advice is no longer always accepted as infallible. Their new patients are less passive, more curious, involved, informed. Gradually, people are beginning to accept the burden of preserving health by exercising, eating more carefully, and discussing the state of the environment with an eye toward its possible effects on them. In larger numbers, they are pushing for the promotion of health, rather than the treatment of disease. No-smoking areas in restaurants and airplanes and the 20 million Americans who claim membership in one or more environmental groups attest to this powerful change; so does the recent founding of the Federal Office of Health Promotion and Disease Prevention within the National Institute of Medicine; so do the books, conferences, seminars, and courses that are dedicated to health education, all trying to increase people's understanding of where their own long-range interests lie and to give them an incentive to work harder to stay healthy. And so do the efforts of millions of people who are conforming to their bodies' requirements by changing their patterns of living whenever they are clearly harmful.

We are experiencing a growth in the emphasis on the personal, relative value of each person's state of well-being. We are witnessing the emergence of the "it-works-for-me" philosophy, as individuals begin to seek the facts about health as it relates to them and learn to respond to those discoveries.

The trend is surfacing on the professional side of medicine as well. Older medical disciplines that emphasized cures are now supported by a dozen newer ones, such as predictive, constitutional, and orthomolecular medicine, all of which tailor their efforts to the individual, and public and environmental health, which work toward the well-being of populations. Even psychiatrists are coming around to a more holistic, balanced view of what constitutes health.

Nevertheless, the changes, while broad, are proceeding very slowly. The idea of working constantly to maintain health (rather than reacting to reinstate it) is difficult for a society accustomed to the "quick-fix" to accept. As a result, even as we inch toward our theoretically ideal life spans of about 90, altogether too many people still die after prolonged suffering and years of dramatically decreased physical activity, after undergoing the psychologically debilitating switch from playing an active, contributory role in society to accepting a passive, disabled existence. Unless our view of the role of medicine changes more quickly, many of us will suffer similar fates.

Clearly, the solutions that science has worked out for the killers of the past are no longer the answer. Chronic diseases tend to accumulate their environmental insults over our entire lives: Plaque in our arteries does not form overnight; most types of arthritis do not develop in a week; and cancer often appears to develop only after someone has been exposed many times to small quantities of carcinogenic agents. Intervening in the course of full-blown diseases like these cannot easily reverse the degenerative processes that often develop over a period of years.

But if prevention is so obviously the preferred route for better health, why are we having so much trouble incorporating it into our lives? Why is it so difficult for us to accept a course of medicine that would lower medical costs, shorten hospital stays, and provide a better quality of health care and health for most citizens? The answer can be found at both ends of the health care system—in the actions of both the providers of care and the recipients.

The providers—our physicians—have been taught to think in terms of treating existing diseases, of rectifying damage that has already occurred. The infectious model of disease that they learned in medical school was based on the assumption that most diseases are acute—and that acute problems require acute care.

Their task has not been made any easier by society's attitudes. People generally consult physicians only when sickness or disability has already struck. As a result, physicians' roles are often defined by the requests made of them; they have always been expected to respond to a problem after the fact.

This attitude has been encouraged by our system of medical education. Medicine's role was to cure disease, and students were taught to do just that. Until recently, courses that could help reorient medical students' thinking were seldom included in the medical curriculum. Chief among these is the subject of human genetics. In a survey of 103 American medical schools, Barton Childs and his colleagues found that more than a quarter still do not require courses in genetics; and those schools that do vary considerably in the time they devote to the subject—from 6 to 54 hours over a four-year period. Furthermore, less than half the medical schools surveyed offered one or more courses in genetics in their continuing education programs, which makes it difficult for practicing physicians to learn of developments in the field. Still, at least the schools are moving in the right direction. Just a decade ago, the number of schools with mandatory genetics courses was only two-thirds of what it is today.

The burden, however, does not rest solely on physicians. For every person who works hard to stay healthy, several make no effort at all. Most smokers, for instance, recognize the destructive nature of their habits. They see the flat statement, "Warning: The Surgeon General Has Determined That Cigarette Smoking Is Dangerous To Your Health," every time they pull out a pack. They have heard the statistics of the suffering smoking can cause. And yet, knowing the odds against them, they continue to smoke. Why? Because many rationalize that statistics are nothing more than numbers. Out of 53 million smokers in this country, only 90,000—two-tenths of 1 percent—get lung cancer each year. Smokers look at these odds and say with some conviction, *it won't happen to me*.

There are two points of view about that attitude. The first, expressed by a substantial chunk of the medical community, goes like this: "They'll never quit. People who still smoke are simply unwilling or unable to face the facts. They will find a way to explain away their habit no matter what the odds."

The second point of view takes into account the potential of genetic prophecy. It assumes that the difference between those who continue to smoke and those who quit is, in many cases, simply a question of tolerance, a threshold which either has or has not been reached. For those who have stopped smoking, the health equation has tipped in the direction of safety. For at least some of those who haven't, the odds of illness still seem remote. To them, the risk of contracting a chronic illness is not yet as powerful a message as the need for a cigarette. And they will continue to smoke until the balance shifts.

Genetic markers for lung cancer, bladder cancer, emphysema, and heart disease could change that equation. They could, in effect, point a finger, singling out those who face the greatest risk. They could personalize those statistics, until the vague numerical value of two-tenths of one percent becomes: You, you, and you.

A Question of Risk

With prophecy, statistics take on new meaning. Estimates based on the *personal* probability of harm allow each of us to decide what the numbers mean to us, and not merely to the population at large.

Nevertheless, it is almost impossible to understand what some of the numbers really imply. Try, for instance, to visualize the import of one chance in ten thousand. And the larger the number gets, the more meaningless it becomes. But if we take those numbers and speak of them in ordinary terms, they become more understandable: Your odds of flipping a coin once and getting heads are one in two. Your odds of flipping a coin seven times and getting seven straight heads are less than one in a hundred.

The risks of contracting a disease fall into a similar pattern. Suppose you are asked to select one marble—your chance of getting ill—from a bowl full of marbles. Those signifying illness are black; the others are white. If your odds of becoming ill are one in one hundred, you would have set before you a bowl with 99 white marbles and a single black one.

If genetic factors were uniformly distributed throughout the population—that is, if everyone had the same chance of contracting a disease—each person would receive the same ratio of black and white marbles. But genetic factors create different risks for different people, and each of our bowls contains a different ratio. Our willingness to play this particular game might change if we were handed, say, a bowl with fifty black and fifty white marbles instead of one with only a single black.

Then again, perhaps it would not. We are always taking risks, even though we are constantly trying to improve the odds for our survival. Every time we get in a car, we are making a small decision to risk the chance of an accident in exchange for the benefits of getting where we are going more quickly. We know that the brakes might fail, the tires might blow, the axle might shatter. Yet we drive anyway, because the certain benefits of using the car outweigh the possible hazards. But what if the possible becomes probable? Will we drive off as readily in a car with bald tires, or one that leaks brake fluid?

So it is with genetic information in medicine. We have an opportunity to learn something more about what we are risking and what the trade-offs are going to be. We can learn whether the odds are against us or in our favor. And we can acknowledge that, in terms of susceptibility, some of us are more equal, and better off, than others. The decision is still ours to make. But now it can be made with new information. Genetic prophecy can help tell us what we might safely enjoy and what we might want to avoid.

Prevention—The Ultimate Cure

Prevention is a long-term commitment that requires constant vigilance. It is more difficult to practice than therapeutic medicine, mainly because while therapy is less effective and more costly, it takes no effort in our day-to-day lives.

These two opposing views of health represent a conflict between living for the present and securing some small guarantee for the future. For most people, it is easier to live for today—a view reflected in the often quoted ancient Epicurean philosophy: "Eat, drink, and be merry, for tomorrow we die." The problem with pursuing that line of thinking is that it is, essentially, a self-fulfilling prophecy. Refusing to take reasonable precautions in the face of obvious dangers simply makes the consequences that much more probable.

Trying to prevent disease before it occurs is like taking out insurance; it entails making small sacrifices today to increase the odds of protection in the future. As always, insurance costs. If it requires that you eat an apple a day—or swallow a vitamin or avoid tropical climates in midsummer—its sacrifices may be simple to accept. But if it demands wholesale changes in a style of living, agreeing to the insurance offered by genetic prophecy becomes a matter of evaluating a trade-off, on deciding whether or not the sacrifices of today are worth the relief promised in the future. If the rewards seem large, agreeing to some deprivation is easier. If they are small, abstaining from something enjoyable becomes that much more difficult.

Genes help define what is at stake by determining who in

the population is at greatest risk and who will suffer less under specific environmental conditions. Along with the other elements that enhance or reduce our susceptibility and resistance to certain diseases—factors like age, nutrition, and lifestyle—genes provide the foundations for a formidable medical arsenal that improves the accuracy of our ability to predict, for each individual, the chances of becoming ill. Indeed, they are potential tools for the insurance industry, among others, since predicting life expectancy and illness—and determining insurance rates on the basis of those predictions—is the key to that industry's success.

Together, prevention and prophecy comprise a critical element in our medical future. As we begin to recognize the changes in the nature of the diseases we face, and as we assume more active responsibility for maintaining our states of health, we are learning to control the factors that predispose us to illness and to design better approaches to prevention. Prevention acts as a catalyst for prophecy; and prophecy enhances our ability to prevent. Or, as Barton Childs has pointed out: "Little by little, a fundamental question in medicine will take on a new form. Instead of asking, 'What is the cause of such and such a disease and how do we treat it?' we will ask, 'What are the reasons why *this* person has such and such a disease, or is predisposed to it, and how can we treat him or help him to manage his life so as to avoid it?'"

PART II
HEALTH

3

OF MICE AND MEN

In the fields of observation, chance favors only
the mind that is prepared.
LOUIS PASTEUR

About twenty years ago, Frank Lilly, then a researcher at Memorial Sloan-Kettering Cancer Center in New York City, sat in his laboratory waiting for two strains of mice to mate. One strain—a speckled brown mouse called C3H—was extremely susceptible to a virus that causes mouse leukemia. The other strain—a C57 black mouse—was resistant to the same virus. Lilly had designed an experiment to uncover the essential difference between the two strains, the difference that protected one from the disease and made the other vulnerable.

The trail leading to Lilly's laboratory was loaded with coincidence. It began in the 1930s, when Peter Gorer, a researcher at Guy's Hospital in London, England, set out to find the mouse equivalents of the ABO blood groups that were known to exist in humans. Groups as basic as those would probably exist in all mammals, Gorer reasoned. And if mice turned up with similar groups, experiments that could not be performed on human beings could be attempted with them.

During his experiments, Gorer came across a previously undiscovered system of mouse antigens—tiny molecules that sit on the surface of cells, governing the production of antibodies, the proteins that help the body fight disease. He named it the H2 system, and promptly found that, since he had no way to evaluate the H2 system's significance, his discovery was greeted with nearly universal apathy.

39

The H2 system remained mired in obscurity for decades. And then, by chance, a clue surfaced in a seemingly unrelated body of research. Ludwik Gross, a scientist at the Veteran's Hospital in New York, had been searching for the cause of leukemia in mice. In 1957, he found one: a tiny squiggle of DNA wrapped in a protein coat, a virus that infected the cells of mouse bone marrow and triggered the disease. Gross knew nothing of the H2 system; genetics was not his field. But his published findings included a list of the strains of mice he had used in his experiments. In England, Peter Gorer read Gross' article. He noted a strange circumstance: Each of the four strains that Gross had used in his leukemia experiments carried the same H2 type, the same H2 antigen out of a possible list of dozens. Was it, he wondered, mere coincidence? Or could it be more, an indication that the H2 system had some effect upon the occurrence of leukemia?

Gorer died before he could satisfy his curiosity, but he had passed his idea on to others. And so, in a quiet laboratory in the middle of New York City, Frank Lilly was waiting for an answer.

The experiment Lilly had designed was simple. The two strains of mice he had selected—the speckled brown and the black—had been inbred through scores of generations until nearly all their heterogeneous traits had been erased. For all practical purposes, both strains were genetically pure, with each mouse virtually identical to all the others in its strain. The speckled brown mice all carried the same H2 antigen that the mice in Gross' leukemia tests had carried; the black mice did not. Lilly crossbred the two strains. Then he took the progeny and mated them again, this time with the original parent strains. As the new litters of between five and fifteen mice were born, he injected every baby mouse with the mouse leukemia virus. A month later he drew blood from each mouse to test for the presence or absence of the H2 antigen. A month after that, he checked them all to determine which had contracted leukemia.

The results were just what Peter Gorer had guessed. Each mouse had one of the two different H2 antigens originally carried by the two different strains. But 95 percent of the mice carrying the same H2 antigen as the mice in Gross' strains contracted leukemia; and only 50 percent of the others came down with the disease.

To check his findings, Lilly performed the same experiment, checking the other known differences (such as color) between the two strains. Not a single one showed any correlation with the incidence of leukemia.

It was an extraordinary finding. For the first time, science had come up with a mechanism for pinpointing which animals were susceptible to a type of cancer and which were not, and the marker of susceptibility turned out to be genetically controlled. Lilly, now chairman of the Department of Genetics at the Albert Einstein College of Medicine in the Bronx, described the quality of the moment: "It was a small revelation. Until then, we knew only two things about the H2 system: that it governed molecules that appeared on cell surfaces, and that it was very complex. Now we had learned that, in some way, the H2 system was *important* as well."

A finding like that might sound minor, but it is not. Basic scientific investigation is fueled by just these kinds of discoveries. For although science purports to search for a basic physical understanding of things, it is not some mindless automaton, marching ever onward toward the truth. It is conducted by people; it is paid for by people; and people make the decisions about what kinds of research should be done. At times, the politics that go into scientific decision making are all-important. That the H2 system was complicated and interesting was not enough to grab people's attention; that the system might be an important element in triggering disease, and might generate expensive, revelatory research projects, was. Partly because of Lilly's findings, some geneticists intensified their studies of the H2 system in mice, while others turned their attention to its counterpart, the HLA (human leukocyte antigen) system in humans.

HLA antigens were studied originally because of their importance in the surgical transplantation of organs from person to person. In the early 1960s, scientists discovered that the antigens were a critical element in the body's decision to accept or reject a new kidney or heart in the delicate period just after surgery. The first HLA antigen was discovered in 1958 by Jean Dausset of France. By 1967, only five others had been isolated. But as their scientific importance and clinical significance increased, the discoveries multiplied. In 1970, the number had reached eleven; some 10 years later, the count stood at 92.

The antigens themselves constitute a kind of biological

signature for the body, a group of molecules that makes it possible to distinguish one individual from another on the cellular level. They are the direct products of a tiny group of genes found on chromosome number six in every human cell. After they are produced, they take up residence on the cell's surface. There they function as an important part of the body's identification system, a set of highly visible I.D. cards that are carried by virtually every cell in the body.

The system of identification is one of the most crucial elements in the body's defense against disease. The system has two main components: First, the *white blood cells* are armies of microscopic soldiers that circulate constantly through the blood vessels and attack and destroy anything they cannot recognize as belonging to the body. Next come tiny molecules, called *antibodies*, that attach themselves to any cell not carrying the proper identification, marking it as something that the immune system should destroy. White blood cells and antibodies check the identification of every cell they find. Together, they form a highly efficient monitoring system that makes it difficult for foreign material to enter the body and disrupt its function. They defend the body against invasion and infection by everything from bacteria and viruses to fungi.

The HLA antigens identify the cells on which they sit as card-carrying members that belong in the body. That is why those cells are left alone. Except in the most unusual and self-destructive circumstances, the immune system is scrupulous in avoiding acts of violence against the body's own cells.

HLA antigens come in a variety of shapes and sizes. They are divided into five separate and distinct sets (A, B, C, D, and DR); each of us inherits a total of ten antigens from our parents: two from each set, one per set from each parent. We therefore can end up with as many as ten different antigens, although within any one set, the two antigens can be identical if both parents happen to pass on the same one.

Scientists have not yet worked out mathematically the number of possible combinations of antigens there might be, mainly because they are not certain that they have managed to identify all the antigens. But the 92 that have already been discovered offer an astonishing degree of variety: When all the combinations among the five sets of antigens are multiplied together, the total number of possible permutations reaches into the hundreds of millions. Mathematically, it is unlikely that any one person on earth is identical in his or her

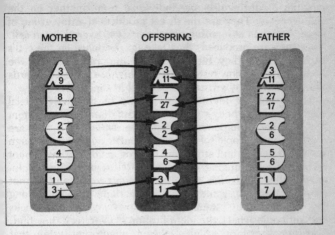

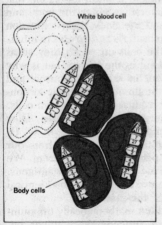

2. **The HLA system.** HLA antigens confer a singular pattern of identification used by the body's immune system. (A) Each parent donates five HLA antigens—one from each of the five groups—to a child. (B) The white blood cells, part of the immune system, identify body cells by their HLA types and leave them alone. (C) White blood cells recognize the foreign nature of cells without the correct HLA type and attack them.

HLA type to more than a dozen or so people. (Because they carry the same sets of chromosomes, identical twins have identical sets of antigens. That is why transplantation between them seldom results in the recipient rejecting the new organ. For those of us who were born without a readily available set of spare parts, surgeons rely on tests of HLA type to locate people with donor organs that match the type of the recipient as closely as possible.)

The Coming of HLA

As research into the HLA system continues, its impact on medical practice increases. Already it has managed to make transplantation surgery simpler and safer. In the future, we may learn to manipulate the body's immune system so that its normal response to tissue that it recognizes as foreign can be suppressed, and transplants can become easier still. We may also learn how epidemics begin, and why one person may have the ability to identify a bacterial or viral invasion and fight it while another does not.

But perhaps more important, the nature of HLA antigens as the body's I.D. cards offers us the opportunity to uncover our personal susceptibility and resistance to scores of diseases.

The possibilities were clear from the beginning. When HLA antigens were first discovered and were found to be the human counterparts of H2 antigens in mice, researchers began to scour populations for correlations between HLA antigens and disease that were similar to those discovered in mice. In France, for instance, people with leukemia were HLA-typed in an experiment to find the human analogue to Ludwik Gross' mouse antigen. But the results were generally negative. Only a decade ago, Jean Dausset could still confess that there was still no "convincing evidence for a correlation between the HLA system and susceptibility to disease in man."

Nevertheless, as the art of surgical transplantation spread, physicians continued to test for HLA types. Time after time, they found specific HLA correlations to specific diseases. They published their discoveries, and the momentum began to build. Investigators began again to examine the HLA types of entire populations of patients, all with the same disease.

And this time they found correlations. The early ones were weak. But spurred on by animal studies that implied a powerful link between HLA and disease, they kept trying. They based their studies on two premises: If an antigen appeared more frequently in a disease group than in the general population, it quite possibly meant that someone with that antigen was more likely to contract the disease than someone without it; conversely, if an antigen appeared *less* frequently, those carrying it might actually be more *resistant* than others. In 1970, the first significant correlation had been confirmed: A link was discovered between HLA and Hodgkin's disease, or cancer of the lymphatic system.

In recent years, activity in the field has run riot, with dramatic successes and still more important promises for the future. Today, HLA antigens are accepted as genetic markers for more than eighty different diseases, with tens more being discovered every year (Table 1).

Perhaps the most significant HLA association yet discovered exists between an antigen from the B group (called HLA-B27) and ankylosing spondylitis, an arthritic condition of the spine, also called "bamboo spine" because of the characteristic look of the vertebrae after they have begun to fuse. Ankylosing spondylitis and HLA-B27 also offer a perfect example of how a genetic marker can be used to predict *predisposition* to a disease even when the correlation between disease and marker is not completely understood.

Consider these facts:

- Ankylosing spondylitis occurs in about 0.4 percent of the population, or about one million Americans, mostly young men.
- HLA-B27 appears in about 8 percent of the general population.
- HLA-B27 also appears in an astonishing 95 percent of those who come down with ankylosing spondylitis.
- About one out of every sixteen people with HLA-B27 comes down with the disease. If the figures are restricted to young men instead of the entire population, the ratio drops to about one in four.

Studies have shown that someone who carries the HLA-B27 antigen is as much as 175 times more likely to develop the disease than someone without it, an enormous increase in

TABLE 1

DISEASES ASSOCIATED WITH HLA GENETIC MARKERS

You Are More Likely To Get:	If You Have These HLA Antigens:
ALLERGIES	
Hay fever (ragweed pollenosis)	A1-B8, A2-B12
Asthma	Bw6, A1-B8
Hypersensitivity pneumonitis (sensitivity to inhaled organic dust)	B40
CARDIOVASCULAR DISEASES	
Hypertension	B18
Coronary artery disease	Bw21
Arterial occlusive disease (thromboangitis obliterans)	A9, B40
Takayasu's disease (pulseless disease)	B5, B40
Mitral valve prolapse	Bw35-A3
IMMUNE SYSTEM DISEASES	
Primary immunodeficiency diseases (difficulties in mounting an immune response)	A2
Ataxia telangiectasia (eye and skin disorders; abnormal balance)	B17
CONNECTIVE TISSUE DISORDERS	
Systemic lupus erythematosus	B8
Sjögren's syndrome (a combination of arthritis and absence of tears)	B8, B37
DERMATOLOGY	
Psoriasis vulgaris	B13, B17, B37, Bw16, Cw6
Dermatitis herpetiformis (skin disorder frequently associated with degeneration of parts of bowel)	B8, Dw3

Pemphigus vulgaris (scaling skin disorder)   A10

Keloids and thick scarring   B14, Bw16

Recurrent oral ulcerations   A2-B12

Behçet's disease (chronic, relapsing oral and genital ulcerations, arthritis, etc.)   B5

Atopic dermatitis   A3, A9, Bw35

ENDOCRINOLOGY

Juvenile diabetes mellitus   B8, B15

Thyrotoxicosis (Graves' disease; disease of the thyroid gland)   B8

Subacute thyroiditis (inflammation of the thyroid gland)   Bw35

Addison's disease of unknown origin (disease of adrenal glands leading to disorders of mineral balance)   B8

Adrenal gland hyperfunctioning   A1, B8

GASTROENTEROLOGY

Chronic active hepatitis   B8

Carriers of hepatitis virus particles   Bw41

Biliary cirrhosis   B15, Bw35

Alcohol liver disease (with cirrhosis)   B8

Hemochromatosis (abnormally high iron levels leading to skin pigmentation, enlarged liver, heart failure, etc.)   A3

Chronic pancreatitis   A1, Bw40

Cystic fibrosis   B5-B18

Gluten-sensitive enteropathy (injury to the gut from ingestion of a food protein, gluten)   B8

Acute appendicitis   B12

Pernicious anemia   B7

INFECTIOUS DISEASES AND RESPONSES TO IMMUNIZATION

Recurrent infection with herpes virus	A1
Infectious mononucleosis	A10
Meningitis and epiglottitis from hemophilus influenza infection	B17, A28
Hypersensitivity to mumps, staphylococcus, and candida	Familial
Leprosy	Familial
Urethritis from gonorrhea	A29
Low vaccination reponse to smallpox (vaccinia inoculation)	Cw3

MALIGNANCY

Testicular teratocarcinoma (cancer of testes)	Dw7
Aplastic anemia	B12
Carcinoma of the cervix	B15
Carcinoma of the kidney	B17
Carcinoma of the rectum	A9
Bladder cancer (transitional cell carcinoma)	B5, Cw4
Hodgkin's disease	A1

NEUROLOGY AND PSYCHIATRY

Multiple sclerosis	B7, Dw2
Optic neuritis (inflammation of the optic nerve)	Dw2
Myasthenia gravis	B8
Motor neuron disease (degeneration of nerves in certain parts of the brain stem and spinal cord)	A2, A28
Paralytic dementia (insanity) following syphilis	B18
Schizophrenia and manic depressive disorders	B17, A28

OPHTHALMOLOGIC DISEASES

Acute anterior uveitis	B27

| Primary open angle glaucoma | B12, B7 |
| Ocular histoplasmosis (fungal infection of the eye) | B7, DRw2 |

PULMONARY DISEASES

| Asbestosis | B27 |
| Farmer's lung | B8 |

RENAL DISEASES

Polycystic kidney disease	B5
Nephrotic syndrome (steroid-responsive; protein in urine, generalized edema)	B12
Vesicoureteric reflux (reversed flow of urine from bladder, leading to infection)	A3
Analgesic abuse nephropathy (kidney damage from analgesic abuse)	A3
Glomerulonephritis after streptococcal infection	Aw19, B12
Familial renal cell carcinoma	Familial

RHEUMATOLOGY

Ankylosing spondylitis	B27
Reiter's disease	B27
Postinfectious arthropathies (joint inflammations after infections)	B27
Psoriatic arthritis (psoriasis appearing in 5-10% of patients with arthritis)	B27, B13, Bx17, Bw38
Juvenile chronic polyarthritis	B27
Frozen shoulder (or periarthritis of shoulder)	B27

MISCELLANEOUS DISEASES

Hereditary hemorrhagic telangiectasia	Familial
Preeclampsia	A1
Normotensive hypertrophic cardiomyopathy	B12 (Caucasians) B5 (blacks)

risk. The correlation between the antigen and the disease explains, among other things, why so few African blacks come down with ankylosing spondylitis, since the B27 antigen rarely appears in that population.

Until recently, however, researchers had not been able to put together a comprehensive set of factors that triggered diseases like ankylosing spondylitis—that is, a list of the environmental *and* the genetic components that, together, could cause disease. But now, thanks to some dogged detective work, some of those mysteries are being solved.

In 1962, the cruiser U.S.S. *Little Rock* was on a tour of the Far East. At its final port of call, six months before it was to return to the United States, the crew decided to hold a picnic.

Because infectious diseases were endemic throughout the Orient, the kitchen staff took elaborate precautions with the foods being served. They carefully wrapped and preserved the meats, washed the vegetables thoroughly, and sliced the bread only as it was being served.

Within a few hours of the picnic, the ship weighed anchor. Eighteen hours later, as case after case of dysentery came stumbling into the infirmary, it became clear that all was not well.

A little digging by the medical staff uncovered what had happened. Despite the kitchen's precautions, two cooks had already come down with dysentery some hours before the picnic. Because they did not want to be hospitalized and lose their promised shore leave, they concealed their illnesses. Instead, as they continued to prepare the food, they made quick dashes to the bathroom. They were so concerned about being caught that they didn't even stop to wash their hands.

Within a few days, 602 of the crew of 1,276 (almost half) had come down with the disease. And while nobody became critically ill, within two weeks ten out of the 600 had contracted Reiter's syndrome, an uncommon type of arthritis that occurs in conjunction with inflammation of the eyes and urethra.

The appearance of so many cases at once had to be more than coincidence. The Navy reported an average of 36 cases of Reiter's syndrome every year among its population of about 904,000. At that rate, the U.S.S. *Little Rock* should have had one case every two years. The odds of ten cases appearing in sixteen days by chance were calculated at 4.1 X 10^{58} to one—an astronomical figure. The only thing the men had in common was the picnic. The obvious conclusion was that the bacteria involved—*Shigella*—somehow triggered the onset of the arthritis.

These findings were duly reported and forgotten until a decade later, when two Stanford University immunogeneticists, Andrei Calin and James Fries, began to wonder exactly what it was that caused only those ten to come down with arthritis while sparing the other 592. By then it was known that HLA-B27—the same antigen that is linked to ankylosing spondylitis—had some connection with Reiter's syndrome. Calin and Fries were curious: Could the men who came down with arthritis have had a predisposing genetic factor?

Information on the whereabouts of the *Little Rock*'s crew was either classified or not available, so Calin and Fries had to track them down, one by one. Ultimately they found eight of the ten who had developed Reiter's syndrome. When they tested each man's blood for HLA antigens, they discovered that seven of the eight carried the B27 antigen. Furthermore, the one man who did not carry the B27 antigen had had the mildest symptoms and was now, thirteen years later, symptom-free; all the others were still suffering from arthritis in their knees, ankles, and wrists, blurred vision (and, in one case, blindness in one eye), and chronic urethritis.

Clearly, the onset of bacterial dysentery had led to arthritis among those with the B27 antigen. An infection with no obvious resemblance to the chronic disease that followed had, in some way, triggered the disease. The propensity for arthritis was somehow activated by the appearance of the bacterial infection.

The case of the U.S.S. *Little Rock* led Fries and Calin to several conclusions. They pointed out the near certainty that HLA-B27 and bacterial dysentery are both instrumental in the onset of Reiter's syndrome. But because not every sailor with B27 and dysentery came down with arthritis (they calculated that 24 to 42 members of the crew had had both), and because one sailor without B27 did, other genetic factors must also be involved. Since then, the search for those other factors has continued.

HLA and Prediction

Knowing that somebody is at risk for a disease is, of course, not the same thing as making a firm diagnosis ahead of time. Although someone with HLA-B27 and bacterial dysentery seems to have about one chance in three of contracting certain kinds of arthritis, a physician might find it difficult to warn a patient before the disease strikes, especially if the problem in question is one that he cannot do anything about. Furthermore, a disease like ankylosing spondylitis cannot, and should not, be diagnosed without an x-ray that confirms the existence of the characteristic bony deformities of the vertebrae. As a result, some physicians maintain that the presence or absence of HLA-B27 is meaningless for determining whether or not a disease will occur in any one patient; while it might give a strong clue as to who is at risk, the antigen alone does not signal the onset of disease. Unless the entire complex of factors surrounding the disease is known, they claim, the presence of the antigen is a mere statistical marker. Using it as a genetic marker might cause certain people to be classified as "B27 cripples," even when nobody can say for certain that they will become ill.

That claim, as far as it goes, is accurate. Until all the environmental and genetic factors that blend to trigger ankylosing spondylitis are discovered, an accurate, confirmed diagnosis can only be made when the disease is present. But the claim ignores two critical areas of medical practice that might be affected by the marker: Those at risk might be monitored more closely, so that the onset of the disease can be discovered earlier; and the marker might have other, unforeseen uses beyond the mere diagnosis of disease.

One such use has already been found. Stanley Hoppenfeld, an orthopedic surgeon at Albert Einstein College of Medicine who specializes in scoliosis—the lateral curvature of the spine—came across a young male patient who had both scoliosis and the B27 antigen. The presence of the antigen gave him the chance to try an entirely new and natural course of treatment.

Today, scoliosis can be treated in two basic ways: by a massive surgical procedure that fuses up to 17 of the 28 spinal vertebrae; and by bracing, in which the patient stays strapped into a Rube Goldberg contraption day and night from the time the diagnosis is made until spinal growth has stopped at about age 16. For this particular patient, however, bracing was not an option; he was 21—his spinal growth had ended years before. And his curve was getting worse.

The B27 antigen, however, provided a third option. Hoppenfeld braced the patient anyway, hoping that the HLA antigen signaled the possibility that ankylosing spondylitis— the natural fusing of the spine—would develop. If it didn't, he reasoned, the bracing would, at the very least, prevent the curve from getting worse; and he would still have the option of operating to fuse the spine artificially at a later date.

As the months passed, it became clear that Hoppenfeld's analysis had been correct. Within a year, the patient's spine had begun to fuse. Within 18 months the fusion was so strong that the brace could be removed. Hoppenfeld never had to operate. An expensive, painful, sometimes dangerous surgical procedure had been avoided. And a genetic marker had paved the way.

The link between ankylosing spondylitis and HLA-B27 is only one of the many connections that have been discovered between HLA and disease. Among the more important:

- Dermatologists have established that one common form of psoriasis is found about five times more frequently in people carrying HLA antigen Cw6. American Indians, who, as a population, do not have the gene for Cw6, do not suffer from psoriasis.

- Gastroenterologists have found that people with the antigen HLA-B8 are three times as susceptible to chronic active hepatitis as those without it. They also have a ninefold risk of developing a sensitivity to wheat gluten, a protein that can cause severe intestinal disturbances.

- Myasthenia gravis, to which Aristotle Onassis succumbed, can be predicted by double genetic markers: both HLA antigens and the sex chromosomes play a role in defining predisposition. Although female Caucasians with HLA-B8 carry a twelvefold risk of developing the disease, B8 has not been shown to predispose males to the disease.

- Bladder cancer is a multifactorial disease. The chances that it might strike are enhanced by smoking, drinking coffee, or working in the dye and rubber industries. But genetic factors may also increase the risk. People with blood group A have a slightly larger chance of coming down with bladder cancer, as do people with a combination of HLA-B5 and HLA-Cw4. For those with both genetic markers, however, the risk is appreciably greater: according to studies performed at the Durham Hospital in England, someone with blood group A and the two HLA antigens has a fifteen times greater chance of contracting bladder cancer than someone in the general population.

- Depressive disorders not only have general genetic components, but specifically involve HLA genes that may affect behavior. Scientists in Rochester, New York, and Toronto, Canada have found that in families with a history of depressive disorders, affected children and parents share their HLA types far more often than mere chance would dictate. While the role of this HLA-linked susceptibility is not known, its existence is clear. It may be that the genes' effects on the immune system cause changes during embryonic development, leaving susceptible people with the characteristic biochemistry that is associated with depression.

- Some researchers contend that people carrying the combination HLA-A1, HLA-B8 or the combination HLA-A2, HLA-B12 are more sensitive to ragweed pollen. Those with A1-B8 may also be more prone to asthma.

- Scientists have found that the people who live the longest almost invariably have received different HLA-A and HLA-B antigens from each parent. They also do not carry those antigens (like B27 and B8) which are most often associated with disease. Two theories exist to explain this extraordinary finding: either having an identical pair of A and B antigens means that you are more susceptible to certain diseases and therefore live a shorter life; or having unmatched antigens protects you against a wider range of diseases, giving you a better shot at surviving.

Despite these and other antigen-disease linkages, the work with the HLA system is only just beginning. The field itself is less than two decades old; and for the first 10 years or so, nobody could quite figure out why the system even existed. After all, why would the body create cells that were different from everybody else's? Certainly not because it anticipated that a surgeon would some day try to transplant organs.

By now, however, there are some reasonable theories as to why the system evolved and how it works. The favored theory rests on the ability of the antigens to identify the body's own cells. If the body is to recognize something as "foreign," it first must recognize something in its own cells that identifies them as "self." That something is the set of HLA antigens. But the ability to distinguish "foreign" from "self" goes beyond such purely external invaders as viruses and bacteria; it includes variants or mutants of the body's own cells. Many researchers believe that, during the millions of divisions our cells go through in a lifetime, potentially harmful mutants arise as mistakes of replication. Some of them may become cancer cells. Because they are recognized as different, however, these mutant cells are constantly sorted out and destroyed by the monitoring system. According to some theories, cancer can sometimes result from a faulty monitoring system, or even from a normal monitoring system that is simply overwhelmed by too many mutants.

Nevertheless, how the HLA system is related to disease is still largely a mystery. Why should there be a link between a particular HLA antigen and diabetes? Between an HLA antigen and hepatitis? Once an antigen is linked to a disease, the scientific problem is to determine the nature of the connection. A correlation might occur for three reasons:

1. A real causal connection can exist between the presence of an HLA antigen and the development of the disease. The reasons for this connection might vary, but one explanation has received popular acceptance: the theory of molecular mimicry, which holds that so much similarity exists between an invading microbe and a specific HLA antigen that the body doesn't recognize the invader as foreign. As a result, the microbe can establish a niche in the body. Some experiments have suggested that the same antibodies that might react against HLA-B27 also react against

certain known bacteria; a person who has HLA-B27 would not recognize the bacteria as being different and would not make antibodies against them.

2. The HLA antigen might actually give certain people a better chance of *surviving* debilitating diseases. Let's assume that everybody is equally susceptible to a particular disease; but only those without a specific antigen succumb to it. As a result, most of the survivors carry the antigen. Someone making a survey of those with the disease would then find a significant tie-in with the antigen; but the link would signify *resistance* to the effects of the disease, not susceptibility to it.

3. Guilt by association. No relationship of cause and effect exists between a specific antigen and a disease. But there is a strong link between the gene that produces the antigen and the gene responsible for the disease, usually because they sit next to each other on a chromosome. The presence of the antigen therefore may act to signal the existence of the other gene.

This particular relationship between a marker and a disease works best within families, since they pass chromosomes and entire groups of genes intact through the generations. It does not matter what the marking antigen is; if it is closely linked to the dangerous gene and is passed on with it, it can forecast the possibility of disease. In a recent study of juvenile diabetes mellitus, Pablo Rubinstein and his colleagues at the New York Blood Center found that some brothers and sisters of children who have diabetes are more susceptible to the disease than others. If both parents are healthy, the best estimate that can be made without HLA typing is that a brother or sister has about one chance in eight of developing the disease. With HLA typing, however, the risk can be pinpointed far more accurately. Depending on the antigen involved, a second child can be targeted as having approximately one chance in 1,000, one chance in 50, or one chance in two of becoming diabetic.

HLA typing—and other genetic markers—achieve their full potential when the knowledge they offer can actually be used to prevent disease. The chief beneficiaries of Rubinstein's work could be those children who escape some of the early complications of diabetes because those around them have been on guard. A child known to have one chance in two of developing diabetes would certainly be more closely watched than one whose chances were one in 1000.

HLA and the Future

The laboratory test that can delineate HLA types is relatively simple and can be performed in many medical centers throughout the world. It is based on the ability of antibodies to recognize specific HLA antigens and to attack them if they do not belong. A technician draws a sample of blood and mixes portions of it with different antibodies. If the sample that is mixed with an antibody which is known to recognize, say, HLA-B8 is attacked and destroyed by that antibody, then B8 must be one of the antigens in the sample.

Today, at least, the major problem with the testing procedure is the cost of the antibodies themselves; so far they have to be purified from human tissue, a process that is difficult and extremely expensive. But the price is dropping fast. Four years ago, it cost about $150 for a complete analysis of HLA antigens. Now the price has halved, to about $75. The advent of genetic engineering, which allows bacteria to be manipulated so that they can produce human antibodies artificially, promises an even larger drop. In about five years, HLA typing will probably cost no more than most other blood tests.

The other, more important problem perplexing those who work with HLA systems concerns the real clinical value of knowing an individual's HLA type. HLA types that have only a small statistical connection with a particular disease are not yet very useful, mainly because they do not offer the physician a predictive tool that he can act on; HLA antigens with a stronger predictive capacity, on the other hand, may become critical in analyzing and monitoring an individual's health. The bottom line is that the value of an HLA antigen increases as the accuracy of its ability to predict grows and as complementary genetic and environmental factors that influence its usefulness are uncovered.

Few diseases are yet understood to that degree. But HLA typing offers potential in other aspects of disease. The most important seems to be that of predicting the prognosis of a disease even *after* it has struck. HLA-B12 has been tied to the effectiveness of chemotherapy as a cure for leukemia. People with HLA-A1 and HLA-B8 who have Hodgkin's disease have a poorer chance of surviving, statistically, than those with A3 and A11, both of which seem to give a patient better odds of surviving past the critical fifth year of the disease. And in one recent survey, 57 percent of those with cancer of the bronchial tubes who also carried HLA-Aw19 or HLA-B5 were still alive—or even disease-free—two years after the diagnosis, as opposed to 13 percent of those without either marker.

Finally, other uses for HLA antigens are also being discovered. The most interesting connection was made recently by Giphart and D'Amaro at the University Medical Center in Leiden, the Netherlands. The two physicians studied births in 3,900 Dutch families and found that the presence of HLA-B18 was associated with an increase in male offspring. In the B18 families, 252 children were born. Of these, 154 were boys and only 98 were girls, a ratio of 3 to 2. The researchers calculated the odds of that happening by chance of about one in 2,500.

Therefore, while using people's HLA types to predict propensity for diseases is still problematical, its future in medical practice is obvious. Most professionals agree that within the next decade the HLA system will become an indispensible weapon in their fight to control disease. They envision a time when it will be automatic for fetuses still in the womb to be tested for their HLA types, for those at risk to be immunized against the diseases to which they are susceptible, and for parents to be forewarned about the environmental factors that might endanger their children.

4

MARKERS AND CANCER

The genes propose, the environments dispose.

BARTON CHILDS

Ever since the first cigarette was lighted, there have been two sides to every argument about the health hazards of smoking. Nonsmokers dredge up lurid tales of blackened lungs, hacking coughs and slow, painful deaths; smokers always seem to have had an Uncle Harry who smoked ten packs of Camels a day, dying at the age of 85 when he lost control of his Maserati during a high-speed chase. The truth of the matter lies somewhere in between. There are compounds in cigarette smoke that are particularly harmful to some people and not to others. Someone who is resistant to their toxicity is relatively protected; someone who is susceptible is at risk.

Smoking causes lung cancer; of that there is no doubt. The details of the relationship between the two are grisly. Lung cancer strikes nearly 120,000 Americans every year. It is the most frequent cause of cancer deaths among men, and the death rate among women is slowly catching up—mainly because they are now smoking more. In Canada, lung cancer accounted for 7.5 percent of all deaths among men over 34 years of age in 1978, almost double the rate for 1966. During the same period, the death rate among women nearly tripled.

Lung cancer also has a poor rate of cure: only 7 percent of the men and 11 percent of the women who contract it manage to live five years. But it does not strike indiscriminately. According to the American Cancer Society, the major risk factors for lung cancer are:

- Heavy cigarette smoking for people over 50
- Starting smoking as early as 15 years old
- Smoking while working with or near asbestos

Smoking itself causes *80 percent* all lung cancers. But its effects do not stop there. Smoking has been linked to ulcers, bronchitis, emphysema, heart disease, and a half dozen other cancers; it is so dangerous that the Surgeon General has labeled it "the single most important, preventable, environmental factor contributing to illness, disability, and death in the United States."

The single most important environmental factor. Strong stuff, especially from an arm of the federal government not usually given to making flat, unequivocal pronouncements. The implication is that if everybody stopped smoking, early illness associated with certain diseases would decrease significantly and the quality and length of our lives would improve.

But everybody is not going to stop smoking. Some will contemplate the odds against their getting sick and think themselves safe, while others will justify their habit in other ways. What we need is a way to identify which smokers are killing themselves and which are not.

Now the first important genetic marker for lung cancer-called aryl hydrocarbon hydroxylase, or AHH—has been found. AHH may not be the only genetic marker correlated with the disease; it may not even turn out to be the best way for finding out who is susceptible and who isn't. But its significant link to lung cancer has opened the doors to genetic prophecy.

When someone inhales smoke from a cigarette, he deposits various different compounds in his lungs. Some are harmless, water-soluble compounds that dissolve in the bloodstream, are carried to the kidneys, and disappear. But other compounds are not as soluble; if the body left them alone, they would remain in the lungs, accumulating until they reached toxic levels. And so, to protect itself, the body manufactures enzymes that transform these substances into compounds that can be excreted. One group of such substances is known as the polycyclic hydrocarbons. They are present in city smog, charcoal-cooked foods, some shampoos (as coal tars), and pesticides; but they are more abundant in tobacco smoke. AHH is one of the enzymes that acts upon them.

The polycyclic hydrocarbons are known as procarcinogens—

that is, while they themselves are not carcinogenic, they can be converted into compounds that are. In the chain of enzymes working to rid the body of the hydrocarbons, AHH is the one that transforms them into cancer-causing compounds.

The enzymes in the chain are like workers in an assembly line. Each enzyme adds on or chops off a different part of the molecule, getting it in the right shape for disposal. Normally, the activity of the enzymes is coordinated: As soon as one has finished work, the next continues the conversion. But if one enzyme works too well—if it modifies too many molecules too quickly—a surplus can occur at that point. And if the molecules in that surplus can attack sensitive parts of the cell, it begins to suffer.

AHH produces one of the intermediate molecules along the assembly line, a highly reactive compound, known as the ultimate carcinogen, which can destroy the activity or structure of practically any part of a cell with which it comes in contact. Daniel Nebert of the National Institutes of Health has found that cells that are converted into cancer cells tend to bind ultimate carcinogens to their DNA. The more AHH activity a cell has, the more ultimate carcinogens are formed; the more ultimate carcinogens, the more likely it is that the cell will become cancerous.

At least three separate studies have found that the level of AHH in the body is dependent on its genetic type. In fact, the studies have uncovered a specific gene that is responsible for determining how much AHH each person can produce.

One study, conducted by Gottfried Kellermann, now at the University of Wisconsin, found high levels of AHH in 9 percent of the population, intermediate levels in 44 percent, and low levels in 47 percent. Kellermann then tested the levels of AHH in lung cancer patients. There he found the amounts of AHH to be shockingly different. In the 50 cancer patients in his test, 15 (30) percent) had high levels of the enzyme, while only two (4 percent) had low levels. When Kellermann calculated the relative risks for the groups— matching them against the unchanging risk that would exist if AHH had nothing to do with lung cancer—he discovered that someone with a high level of AHH is *36 times* as likely to contract lung cancer as someone with low levels. For those with intermediate levels, the risk turned out to be 16 times as great.

Supporting evidence for Kellermann's findings has come

from a variety of sources. In one study, Rolf Körsgaard and Erik Trell of Norway examined the AHH levels of 102 patients with oral, laryngeal, and bronchogenic cancers. In a normal population of 102 people, 10 people with high levels (9 percent) should appear; Körsgaard and Trell found 39. And where they should have found 47 people with low levels, they found a mere 21. Significantly more people who had high levels of AHH developed cancer than those who had low levels. And the majority of the patients with cancer were also heavy smokers.

A Finnish study by Carl Gahmberg at the University of Helsinki provided further proof. He found high levels of AHH in 39 percent of patients with untreated lung cancer, compared to 15 percent of a healthy control group. He also noted another important difference. Among those who had lung cancer, patients with high AHH levels developed the disease an average of five years earlier than patients with low AHH levels. High levels served as a marker for both *increased* and *earlier* risk of lung cancer.

The implications of these findings are clear: Those who produce lower levels of AHH can be less concerned with the dangers of lung cancer; those with high levels might find themselves with enough incentive to quit smoking. And those stubborn few who continue to smoke despite the odds might, at the very least, agree to frequent medical check-ups so that any appearance of the disease can be caught early enough to give them a shot at surviving.

There are still two drawbacks to the widespread testing of AHH levels. First, no reliable mass screening test has yet been developed. The best test known for AHH requires a liver biopsy, a complicated and sometimes dangerous procedure. The test that might be used in the future—a simple blood test in which the level of AHH in the white blood cells is measured—is not yet dependable, mainly because everything from aspirin and cigarette smoke to other malignancies and the timing of the test affects its accuracy. The blood test may eventually prove useful, but only after all the variables can be taken into account and controlled.

Nevertheless, investigators are already talking about designing "test kits" which can be marketed throughout the country. Richard Kouri of Microbiological Associates in Bethesda, Maryland, is hopeful that the day-to-day variability in the tests for any one person will soon be overcome. His group has developed a method in which a physician takes two or three blood samples over a period of a year, keeps them frozen in liquid nitrogen until all the samples are available, and then tests them simultaneously for AHH activity. Variations that might otherwise confound the test results can be minimized and an average value can be obtained.

The second drawback is that Kellermann's findings are still controversial; some investigators have been unable to duplicate his finding of three distinct groups with different levels of AHH and have managed to isolate only high and low groups. Despite this inconsistency, the significant finding still holds: People with high levels of AHH are found to have lung cancer more often *if they smoke cigarettes* than those who have low levels or don't smoke. A genetic predisposition and an environmental factor are interacting to cause disease. Animal studies confirm these conclusions. The problem, it seems, is not with the theory but with our ability to test for it.

Finding genetic markers like AHH, or any factors associated with disease for that matter, is especially difficult in humans, primarily because we are such poor laboratory specimens. Everything about us mediates against the kinds of clear, simple tests that uncover facts about disease. Unlike mice and rats, we cannot be made genetically pure through breeding; we have comparatively few offspring; and we change our environments and social habits over the years. Most importantly, researchers cannot collect one hundred human babies, inject them with various noxious cultures, confine them to unchanging surroundings, and, at their leisure, assess the results. It is because of these limitations that science has come to accept experimental evidence obtained from other animals as reasonable projections of what will probably occur in man. And it is because of these limitations that drugs and therapies that have been tested in animals before being approved for human use sometimes go awry.

Kellermann's thesis has been tested repeatedly in animals—most often in mice. And it has repeatedly proved true. In one of the more impressive tests, Daniel Nebert of the National Institutes of Health injected a polycyclic hydrocarbon similar to those found in the tar of cigarette smoke directly into the windpipes of two strains of mice, one with high AHH activity, one with low. Fifty-five percent of the mice with high activity developed lung tumors; only 17 percent of the low group were afflicted.

In a parallel experiment, related mice that shared some of their genes were also treated. The group with lower AHH activity had no cancers, whereas 23 percent of the group with high activity (almost one in four) contracted lung cancer. The experiment implied that while AHH activity is an important marker for lung cancer, other genes might modify the degree of susceptibility, and that a particular *combination* of genes is probably necessary before the cancer can develop. If, for instance, a mouse inherits a gene for high AHH activity as well as a gene for high activity in the *next* enzyme in the assembly line, the excess carcinogen will be converted to a harmless substance before it can do any harm. That is why testing for AHH is only an approximation of the risk for cancer. More accurate predictions will become possible when we know the importance of other factors as well; the more markers we can pinpoint, the better our predictions will be.

In the past few years, researchers have been examining the AHH system for its effects on other problems. Perhaps the most frightening experiment, also performed by Daniel Nebert, involved injecting pregnant mice with polycyclic hydrocarbons to determine the effect on the offspring. The results were striking: Embryos with high levels of AHH were stillborn, born with malformations, or reabsorbed by the mother before birth (an indication of a birth defect) between five and twenty times as often as embryos with low levels of AHH. This study, in fact, was the first time that abnormalities in fetuses were predicted on the basis of AHH levels. The implications for human pregnancies are obvious.

Although the liver biopsy test for AHH is not completely suitable for mass screening and the test of white blood cells is not yet reliable, another promising approach is being developed, which measures the levels of AHH indirectly by evaluating the body's ability to break down certain specific drugs. If it turns out to be an accurate measure of AHH production, it may well become a tool of general screening.

The Modern Black Death

Cancer is the number two killer in the United States today. It causes nearly 400,000 deaths each year, only half as many as the annual toll for heart and vascular diseases. Why then are we so preoccupied with cancer?

There are three primary reasons: First, cancer *is* the number one killer among young adults between 25 and 44. It is also the number two killer of those even younger, second only to accidents and suicide. Cancer is therefore a disease that kills people before their time, a fact which makes it seem particularly cruel.

Second, no part of the body is immune from the disease. Cancer can strike the brain, lungs, breasts, genitals, skin, blood, bones—anywhere there are cells. With heart disease, at least, we know where the hammer will fall.

Finally, the course of cancer is protracted and characterized by painful physical symptoms. Its treatment alone, with the nausea of chemotherapy or the baldness and physical weakness from radiation, is worse than many diseases. Throughout the illness, the treatments, even the sporadic remissions, the psychology of cancer hovers over the patient and his family. It thrusts mortality on us; it presents us with the inevitability of death. Cancer seems like a nightmare, a blow from some angry god, leaving us as helpless as our ancestors must have felt when they were confronted by the plague.

But cancer is no longer that scary demon from the dark. Many forms can be treated, especially if they are discovered early enough. Cancer is, in fact, not one disease but a common name for a collection of different diseases characterized by the uncontrolled growth and spread of abnormal cells. In time, these cells can disrupt normal function and cause death. They can also migrate throughout the body, forming new colonies, or tumors, wherever they settle. Even such commonly known cancers as leukemia can actually be any one of several cancers, some of which are more deadly than others.

Thus, cancer is a code word that encompasses an entire group of diseases. Different genes may predispose a person to different cancers. Different environmental agents may trigger the development of different forms of the disease. And different therapies vary in their effectiveness, depending on the type of cancer and the extent to which it has spread. It is as wrong-headed to talk about a "cure" for cancer as it is to believe that a single act of government can alleviate the woes of a complex economy.

But if the causes and possible cures for cancer sound especially vague, they are not. For the past thirty years, investigators have been looking into the mechanisms of cancer—that is, the triggers, predispositions, and gene-environment interactions that lead to it. In fact, they began searching for markers in the first place partly because so many hereditary conditions seemed somehow linked to cancer. And they have managed to come up with some general findings that point directly to genetic prophecy.

Carcinogens. Nobody really knows exactly how cancer develops in the body—how a cell or group of cells begins to go haywire, proliferating out of control. Nevertheless, impressive evidence now suggests that cancer is the result of two separate factors: a cell that is ready to undergo change; and an external element that sets it off. The changes probably begin at the most basic level of the cell, in its genes. The triggering agent, called a carcinogen, can be a chemical, a virus, or radiation.

Experiments have shown that cigarette tars, benzene, vinyl chloride, asbestos, ultraviolet light, x-rays, and certain viruses do not attack the cell as a whole; they damage its genes. They can cause the genes to mutate—to change their code, to produce useless proteins, or to lose their effectiveness. It is probably these alterations in the cell's ability to regulate itself that cause it to go out of control. And it is at that point when cancer begins to take hold.

Carcinogens are everywhere. They are dumped into streams and water supplies. They are spewed into the air. They are injected into food and sprayed on the paper it is wrapped in. They are, in short, an integral part of our industrial society, so pervasive that they manage to initiate about 80 percent of all cancers—with the other 20 percent triggered by natural radiation, viruses, and constitutional abnormalities.

That is not to say, however, that genes are not involved. Even in the most dangerous environment, not everybody comes down with cancer. In the final analysis, cancers are the result of the right environmental trigger attacking the right gene or genes. Both must be present for the disease to occur.

Nearly every cancer illustrates this relationship. One of the most clear-cut cases is retinoblastoma, or cancer of the eye, which is linked directly to a defect on chromosome 13. The chromosomal defect is an excellent marker for the disease; physicians can predict with an accuracy of almost 90 percent who will develop tumors. But what about the other 10 percent? Recent evidence suggests that the disease is triggered by a virus which, in baboons, attaches itself to specific chromosomes in the cells of the eye's retina. If that same virus acts upon humans, it is obviously quite common—common enough to infect nine out of every ten people who are susceptible to it. But those with an intact chromosome 13 can resist it; and the lucky person who carries the chromosomal defect while remaining healthy probably avoids coming in contact with the virus.

Other chromosomal abnormalities that can lead to cancer are as identifiable as the defect on chromosome 13. But most cancers are linked to more subtle genetic susceptibilities. Sometimes, the environment itself can create predispositions in those who would otherwise remain resistant. Parasitic infection can increase some people's susceptibility to bladder cancer, although nobody has yet proved how it happens. And someone who accidentally swallows lye can damage his esophagus so much that it becomes especially vulnerable to the growth of tumors, possibly because cells are more likely to pick up carcinogens when their protective membranes have been damaged.

The Immune System. Even when cancer cells do develop, the body has its own first line of defense. The immune system, which includes the HLA antigens, can recognize cancer cells as foreign, attack them, and eliminate them as threats. In healthy individuals, the system is efficient enough to destroy most abnormal growths. But in those who have weaker or malfunctioning immune systems (or who spend too much time in environments where carcinogens persist), abnormal cells can escape detection or simply overpower the immune response.

In a sense, a strong immune system is a marker for resistance to cancer. Many studies have shown that people with weak systems are far more likely to develop tumors than the general population. For them, the risk of certain blood cancers, for instance, can be as high as one in ten, depending upon the disease and the degree of weakness. Perhaps the best evidence comes from studies of patients who have undergone transplantation surgery. Unless they receive grafts or organs from people with identical tissue types, their immune systems must be suppressed so that they won't automatically reject their new tissue. The risk of cancer soars. In one group of Australian patients, seven out of 51 (almost 15 percent) eventually developed skin tumors after transplantation surgery and treatment with immunosuppressant drugs.

Different genetic markers exist for most kinds of cancer. And while markers of general susceptibility may ultimately prove useful, the most important, like AHH, will identify those at risk for specific diseases. In fact, of the more than 2000 different traits that have been linked to single genes, about 200 are directly associated with an increase of one type or another of cancer. In some, triggering cancer seems to be their only reason for existence.

Breast Cancer

Perhaps no disease frightens more women than cancer of the breast. The disease is common: one out of every sixteen women in the United States will contract it at some point in her life. But the *fear* of breast cancer is even more common: one out of every *three* women *believes* she will contract it. And even though the cure rate is fairly good, contemplating life in a "mutilated" state prevents many women from seeking the early help that could save their lives.

Regular medical screening for early diagnosis of breast cancer has been recommended for years by the American Cancer Society. Most doctors' offices distribute pamphlets that describe how a woman can examine her own breasts regularly for suspicious lumps. And such recent developments as x-ray mammographies, thermal procedures (which measure the increased heat given off by cancerous cells), mechanical palpation machines, and special photography are all designed to detect cancer as early as possible.

There is, of course, an obvious drawback to all these procedures. They all detect tumors which have already started to form; some are neither as accurate nor as safe as they might be. Self-examination, for example, can uncover only the larger tumors or "knots" in the breast tissue. And x-ray mammography has come under attack because regular exposure to irradiation may itself be a cancer-causing factor. Currently, medical practitioners are debating whether annual screening programs should include all women or only those at risk. But who is at risk?

Genetic prophecy can provide at least a partial answer. Breast cancer has long been observed to run in families. On the average, a woman whose family has experienced at least two cases of breast cancer has a lifetime susceptibility of one in six. If those family members should be closely related (her mother and sister, for instance), the odds go up to a very scary one in three. Because of these statistics, a small but growing number of women have agreed to "prophylactic" breast surgery—even though their breasts are still normal, the tissue inside is surgically removed and replaced with silicone plastic.

This is a drastic form of prevention. It is frowned on by many doctors, not only because nobody knows which of these women would have developed cancer, but because cancer can still appear in the remaining tissue. Nevertheless, to some women in the high-risk group, the unpleasantness of the procedure is worth the trouble if it diminishes the risk of the disease.

Identifying women who are truly in greater danger of contracting cancer is a major priority in cancer research today. One possible clue comes from a genetic marker discovered in the oddest of circumstances.

Nicholas Petrakis, a hematologist at the University of California School of Medicine in San Francisco, has had passing interests in genetics and anthropology as well as in his specialty, the scientific study of the blood. In one of his excursions into anthropology, he learned about a characteristic commonly found among Oriental groups: dry ear wax. Caucasians and Africans tend to have wet ear wax; the type any one individual gets is genetically controlled. Intrigued, Petrakis took it upon himself to confirm this interesting detail by examining the ears of American Indians at a local baby clinic. Sure enough, he found that the more Oriental in type the baby was, the greater was the chance that it had dry ear wax.

All of this might have remained an interesting footnote. But Petrakis knew that Asia is known for its low incidence of breast cancer, compared to the rest of the world. Was the connection between dry ear wax and low rates of cancer merely coincidence? During a tip to the Far East, Petrakis stopped off in India to check patients and find out. The correlation stood out: women with wet ear wax were more likely to get breast cancer than those with dry ear wax.

As Petrakis thought about it, he realized that the connection between ear wax and cancer might be more than superficial. Both the ear and the breast have similar glands; both secrete similar fluids. Petrakis wondered whether the ability to secrete these fluids was the same for both areas. So he designed a cup-and-syringe unit that could fit over breasts and suck out fluid; then he recruited 5,000 women to take part in his test. Petrakis found that most Caucasians normally secrete a fluid that can be withdrawn in 10 to 15 seconds of suction; most Chinese and Japanese women secrete at a much slower rate. And most important, the women who secreted breast fluid relatively quickly had wet ear wax.

No scientist makes a finding like this without trying to provide an explanation. Petrakis discovered that breast fluid in general can pick up and concentrate such substances as barbiturates, fatty acids, and compounds from cigarette smoke. It also contains chemicals that are known to cause mutations. These two factors seemed to provide one explanation as to why Caucasian women, whose glands excrete relatively large quantities of glandular fluids, have a higher incidence of breast cancer than Oriental women. Ear wax type just happens to be a clear marker of this genetic propensity—a marker that can be evaluated in a simple test by a physician.

Petrakis' work provided welcome news to most Oriental women, as well as to the 5 percent of Caucasians whose glandular secretions make them more resistant to breast cancer. But the 95 percent of Caucasians with wet ear wax were still in the dark as to how susceptible they actually were. And so Petrakis and his associates began to screen dozens of potential markers—blood groups, HLA types, and enzymes among them. Not a single marker showed up more frequently in breast cancer patients than in normal controls. Perhaps they would have better luck, they reasoned, if they studied families in which breast cancer appeared in clusters.

With Mary-Clair King of the University of California at Berkeley as chief investigator, the team of scientists examined families in which at least three cases of breast cancer could be confirmed in groups of mothers, daughters, and sisters. Eleven families with 426 members were tested for a whole list of markers. And one marker, a variant of a common enzyme, was sifted out.

The marker was called glutamate-pyruvate transaminase, or GPT. Its connection to breast cancer in susceptible families was dramatic: Of those who carry the GPT marker, one in eight develops cancer before age 35, one in two before age 50, and a striking nine out of ten before age 80. Members of these susceptible families who do not carry the GPT marker have no greater risk of cancer than the average woman in the general population.

Although GPT signals a predisposition to breast cancer only within cancer-prone families, it is a useful step toward the discovery of the group of markers that will someday predict the onset of the disease. Within breast-cancer-prone families, we can now determine who is at increased risk. Those women will be advised to pay special attention to methods of early detection; they might be the ones who are most likely to benefit from prompt, and perhaps even prophylactic, surgery.

Skin Cancer

Red hair and sunlight. The combination is one of the most common causes of skin cancer, both experimentally and statistically. The Irish, and others of Celtic origin, are especially susceptible.

Scientists have several theories as to why this is so. Perhaps the most popular is the protective melanin theory, which holds that melanin, a dark pigment, protects the skin against the ultraviolet radiation of the sun. Redheads are usually light-skinned and do not have enough melanin to protect themselves against ultraviolet rays. Their body's reaction, as it forms freckles, is in itself a form of cellular mutation.

Recently, however, another theory has been offered. Investigators at Cornell University Medical College in New York City have isolated the pigment, called phaeomelanin, that is responsible for red hair. It is also found in skin. When test cells were exposed to pure phaeomelanin, nothing unusual happened. But when the same cells were exposed to pigment that had been subjected to ultraviolet light, they began to mutate in the same way that known cancer-inducing chemicals trigger mutations.

Scientists now believe that the phaeomelanin may be a marker for skin cancer. Sunlight may cause the pigment to change so that it becomes capable of causing cancer. While someone born with the pigment can do little to change his or her coloring, avoiding excessive exposure to the sun is one way to minimize the risk of skin cancer.

Several other genetic diseases linked with skin cancer also involve increased sensitivity to radiation. One, called xeroderma pigmentosum (or XP) causes freckling, reddening, blistering and a tendency to scar when the skin is exposed to sunlight. If people with XP are not careful, they usually succumb to multiple skin cancers before they reach the age of 21.

Careful of what? One study showed that XP patients managed to escape skin cancer entirely simply by avoiding exposure to strong sunlight. With that in mind, two doctors at the University of Oregon Health Sciences Center have advised families with XP children to "move en masse to western Oregon and enjoy the endless gray mist and drizzle of our area." Somewhat more seriously, they suggest protective clothing, broad-brimmed hats, special sun screens, and frequent whole body examinations by a dermatologist for those at risk.

Kidney Cancer

Just a few months before the GPT marker for breast cancer was discovered, a marker was found for tumors of the kidneys. It too is significant only in cancer-prone families.

The search for a marker began when a Boston physician found that one of his patients had malignant tumors in both of his kidneys. The finding was unusual: the man was only 37 years old, and less than 2 percent of kidney cancer victims have both kidneys affected. A family history revealed that an aunt also had had kidney cancer; and further digging turned up ten more victims out of the forty members of three generations. Of these, six had cancer in both kidneys.

Cell specimens were taken from the living victims in hopes that a genetic marker might be found. The study paid off: When the cells were examined under a microscope, the doctors could see clearly that a tiny piece of chromosome number 3 had switched places with a similar piece of chromosome 8. The switch took place only in those with the cancer.

The Boston doctors are now keeping close tabs on members of the family who carry the marker. They believe that these people have a 90 percent chance of coming down with kidney cancer, as opposed to the odds of one in 1,000 for the general population. If cancer symptoms *do* show up as expected, they are more likely to be noticed immediately and treated with the speed that so often means the difference between life and death for cancer victims.

A Family Matter

The markers connected with the more common cancers point out one particular genetic connection that the average layman has long understood through observation and physicians have learned from some cold, hard experience: Cancer often runs in families. Some markers that seem to have no particular significance when they are observed in the general population can be used within specific families to point out important differences in susceptibility. The explanation is similar to that offered by Pablo Rubinstein when he and his colleagues discovered the familial links between HLA and diabetes: Families often pass intact collections of genes from one generation to the next; a marker may be associated with a disease, not necessarily because it is actually involved in causing the disease, but because it sits close to a dangerous gene and is passed on with it. In the general population, the neighbor-to-neighbor relationship between the genes may not exist; but in families it can stand out very strongly.

Because of this familial linkage, physicians who are aware of the importance of prediction are beginning to pay far more attention to the family histories they take. No longer does a complete physical examination merely attempt to pinpoint someone's condition at a particular place and time; now it is equally important to find out about heredity, to discover the way genes might run through generations. Markers within families can ease the anxiety of those who might not be at risk and focus attention on those who are actually on the firing line.

Some physicians have learned this lesson the hard way. One prominent pediatrician had spent years in the care of a brilliant internist, whom he considered his family doctor. One day he happened to attend a lecture on new discoveries about cancer of the colon. He learned that the growth of malignant tumors in the colon is almost always preceded by a clear marker: small polyps, benign masses of cells that are easy to locate on the colon's surface. The link between polyps and cancer is almost absolute: 95 percent of those who get polyps contract cancer before they are 40 years old. In addition, the growth of polyps and cancer is tightly tied to inheritance.

The physician left the lecture chilled by two thoughts: His father had died of cancer of the colon. His internist had never asked him about it; nor had he checked his colon for polyps. It took the physician less than 24 hours to find himself another doctor.

In this day of sophisticated medical treatment, no decent doctor should treat a patient without getting a full and specific family history beforehand. And no informed *patient* should allow a doctor to get away with not knowing about potentially life-saving information.

The tendency of cancers to run in families actually goes beyond those specific cancers for which genetic markers may have been found. In general, relatives of people who contract cancer are more likely to get the disease themselves. In the normal population, 30 percent of cancer patients have one close relative who also has cancer; 20 percent have two such relatives; and 7 percent have three or more. Leukemia, the cancer of the bone marrow that seems to attack young children, gives an identical twin odds of one in five if his or her twin has caught the disease. Among other siblings, the odds are smaller; but the connection is clear. It is not unheard of for families to find all their children developing leukemia—a pheomenon that, in the general population, would occur in about one family in a billion if it were by chance.

One family whose records were published by the National Cancer Institute illustrates the extent to which familial tendencies can run. The first patient to be identified as a cancer victim was a woman who had cancer of the cervix. It turned out that her brother had cancer of the colon, her sister had cancer of the breast, and two nephews had contracted a rare and fatal cancer of the blood. Soon after, three of her six children developed leukemia, as did several distant relatives.

The family seemed to develop cancers the way most families pick up common colds. Their record was so extraordinary that investigators decided to test the cells of every member of the immediate family for their tendency to become abnormal (or transform too easily) when they were infected by viruses. John J. Mulvihill, the head of the Clinical Genetics sections at the National Cancer Institute, explained the results of the tests:

> Cells from the father and from two clinically normal twin brothers had normal transformation; high transformability was seen in cells from the mother, a leukemic daughter, and an older normal daughter, who seemed to have been spared from leukemia. *Seven years later, as if predicted by the test, she became the fourth sibling with leukemia*. [italics added]

A Marker for General Susceptibility

Most cancers are joined by a common thread: They develop in response to some kind of abnormality in chromosomes or genes. The more striking chromosomal abnormalities have long been tied to cancer:

- Children born with Down's syndrome, or mongolism, are more prone to leukemia than others.
- Individuals with Klinefelter's syndrome—an extra X chromosome added to the normal XY complement of sex chromosomes—tend to have an increased chance of contracting breast cancer.
- Those missing a piece of chromosome 13 almost invariably develop cancer of the eye.
- People with the so-called Philadelphia chromosome have higher odds of contracting leukemia.

These gross genetic abnormalities are probably just extremes of what happens when any cell is genetically predisposed to cancer.

The interaction between gene and environment is an accepted part of the cancer picture. But how does the environment affect the gene? How does an external insult trigger the unregulated growth and proliferation of the cell?

Scientists now suspect that both normal cells and cells predisposed to cancer suffer similar damage from the environment. But normal cells seem to be able to repair themselves; defective cells cannot. As a direct result of their defect, the abnormal cells begin to mutate.

One of the first clues to the possibility that some cells lack the genetic ability to correct minor disruptions in their DNA sequences appeared in 1975 in Birmingham, England. There, a patient with a rare disease known as ataxia telangiectasia (AT) who was undergoing radiation treatments began to develop acute radiation poisoning. At first, nobody could understand why: the dose he was receiving was no higher than that which normal people accept without difficulty.

His physicians took samples of his cells and examined them. They found that the cells received the same punishment as normal cells undergoing the treatment, but the AT patient's cells could not rebound from the beating. Somehow, they lacked the normal ability of healthy cells to repair the damage caused by environmental stress.

The fragility of certain people's genes and chromosomes has now been firmly connected to their chances of coming down with cancer. Some specific diseases, while rare, seem to stem directly from this fragility. Others, like leukemia, seem to occur with far greater frequency in those who carry at least a partial susceptibility to these so-called chromosome instability syndromes. One rare blood disease, known as Fanconi's anemia, predisposes its victims to infection as well as to death by cancer. But more importantly, close relatives of victims— people who are probably carrying one of the two genes that are needed to trigger the disease itself—also show an increased risk of cancer. Michael Swift, of the University of North Carolina at Chapel Hill, has recorded nine times the number of cancer deaths among younger relatives of Fanconi victims than would be expected in the population at large. His conclusion: the more likely it is that a relative carries a copy of a gene for Fanconi's, the higher the odds that he or she will die of cancer. The chances of contracting leukemia alone seem to be about twelve times as great for these people as they are for the general population. And while the two-gene combination that causes Fanconi's anemia is a rare phenomenon, an estimated three-quarters of a million people in the United States carry a single copy of the gene.

The tests that have been performed on patients with these rarer diseases are now being broadened and extended to those who may be predisposed to other forms of cancer. So far, similar defects have been found in several forms, and scientists are beginning to suspect that the heightened sensitivity of some cells to viruses, radiation, and chemicals may pinpoint them as being predisposed to turning cancerous. The National Cancer Institute's (NCI) viral tests on that family with the extraordinary predisposition to all sorts of cancers is one indication; another is a collaborative effort between Malcolm Patterson and his colleagues at the Chalk River Nuclear Laboratory in Ontario, Canada, and Robert Miller of the NCI. The two groups took skin cell samples from a cancer family with a high incidence of acute myelogenous leukemia, a bone cancer. Then they grew the cells in the laboratory and exposed them to radiation. The samples were coded so that the scientists could not know which were from cancer patients and which belonged to their healthy relatives. They checked each sample for its sensitivity to radiation—for the dose that was needed to kill it. Then they tried to identify the cancer patients from the results. When the code was broken they found that they had been successful: the cells of those with cancer were clearly more sensitive.

Tests like these seem to identify cells that have more difficulty adapting to environmental insults, which is the exact process that many scientists believe leads to cancer. Still another, similar approach is being used by Richard Albertini at the University of Vermont; he has begun to compare the white blood cells of normal people to those of people undergoing chemical treatment for breast cancer. Albertini's test is designed to determine whose cells are more likely to mutate when they come in contact with chemical carcinogens. Because some people who undergo chemotherapy develop other tumors in response to their treatment, Albertini is trying to determine whether the tendency of white blood cells to mutate can be used as a marker for susceptibility.

The three tests for sensitivity—to viruses, radiation, and chemicals—are perhaps the most promising avenues of research toward finding a general marker for cancer. If cell sensitivity proves to be an accurate marker, we may be able to identify those *generally* at risk and monitor them closely for emerging signs of the disease. Daily, we are discovering more carcinogens, more sources of environmental insults, more compounds that we are told to avoid. The possibility that we may be on the verge of a marker for cancers in general is a sign that science is fighting back.

5

ON THE JOB

> . . . it is artificial to question screening in
> terms of an abstract right of equal access to a job.
> One could hardly defend the right of a hemophiliac
> to be employed as a butcher.
>
> BERNARD D. DAVIS

There is no such thing as a risk-free life. Our ancestors battled infidels and starvation; we struggle with carcinogens and the stresses of our fast-paced lives. We cannot expect it to be otherwise.

But we *can* make decisions about the kinds of risks we are willing to face. We can minimize some dangers and make our environments as safe as is practically possible. And we can demand that others do the same.

The choices that we make cut to the very heart of the politics of health. They extend beyond the questions of personal preference that genetic prophecy has raised thus far, beyond simply choosing a way of life that may be healthier or more dangerous, freer or more restricted, than what we had before. They may influence the way we mold society as a whole; we may decide to assume complete personal responsibility for everything that touches us (the politics of libertarianism), or, at the other extreme, we may accept a kind of social paternalism, in which the government monitors and protects us from womb to tomb. The critical issues are: *Who is responsible for deciding how to balance hazard and health? How far does that responsibility extend?*

Nowhere are these issues more relevant than in the workplace. After we accept a job, we find ourselves having less choice and less control over the safety of the work environment than anywhere else. Our dependence on our jobs means that we cannot afford to try to change working environments casually.

The degree of job-related hazard varies. When we decide what we want to do for a living, we naturally take it into account. For people like Evel Knievel, the risk is the job itself; the odds against his surviving a jump over the Snake River in a rocket-powered motorcycle some years ago were a matter of public speculation. A stenographer in the office typing pool, on the other hand, can be fairly sure that life-threatening risk is not a characteristic of the work. In each case, the key to making an informed decision is knowledge about what the risks are and to whom they apply.

But is even that quiet office job really as safe as we think it is? Away from the harsh elements of the outside world, safe from excessive physical exertion, clerical workers certainly *seem* to be protected against hazards. Nevertheless, they, too, face risks. Recent studies suggest that some of the inks used in photocopying machines may be carcinogenic. One casual survey by students at the Cornell University Medical College uncovered the unsettling fact that, of all the surfaces they could think of testing—subway seats, food counters, bathrooms, desks—the most contaminated places in New York City were the mouthpieces of telephone receivers. Carpets are known to be germ jungles: environments that retain dirt and grow rich in the substances on which bacteria thrive. Fluorescent lights provide only a part of the wave-length spectrum available in natural light, a spectrum that some studies have shown to be vital to good health. And chemicals that office workers take for granted have never been adequately tested to determine whether they are, in fact, safe: correction fluids for typewriters, new glue and adhesive preparations, the cleaning and thinning fluids used throughout any proper office.

If offices pose unanswered questions like these, can other work environments—from hospitals to garages, fast-food joints to garbage trucks—be far behind? Obviously, *every* work environment poses its own particular brand of danger.

Most of us, at least, can choose the types of gambles we take. According to the 1980 U. S. Census, Americans can now select from among 23,000 occupations, over two and a half times as many as in 1910. The number is still growing, as are the risks.

Perhaps the greatest risks crop up in factories where, as a matter of course, workers handle substances known to be hazardous. The problems of hazards in the workplace loom so large that the National Institute of Occupational Safety and Health (NIOSH) recently undertook a three-year study to determine which industries are the most likely to trigger cancer. Surprisingly, large chemical companies ranked only twelfth on that list. Companies that produce industrial and scientific instruments topped the charts—apparently because their workers have to handle such carcinogenic chemicals as solder, asbestos, and thallium; next came companies that use nickel, lead, solvents, chromic acid, and asbestos to fabricate metal products. Other industries in the top ten included those building electrical equipment (exposure to lead, mercury, solvents, and solder); transportation equipment (hazardous constituents of plastic, like formaldehyde); machinery (cutting and lubricating oils); the petroleum industry (benzene, naphthalene, and aromatic hydrocarbons); industries making leather products (chrome salts and other chemicals used in tanning); and those building pipeline transportation (petroleum derivatives and welding materials).

Although NIOSH's effort to rank industries was an important first step, its list does not reflect all possible risks. NIOSH looked only for known or highly suspect carcinogens. Industries using chemicals not yet recognized as carcinogens are missing from the list, as are those using materials that might trigger diseases other than cancer.

Nevertheless, identifying cancer-related industries is the first step in taking preventive action against illnesses arising in the workplace. Some people may want to avoid the most hazardous jobs altogether; others may act on the information to lower the risks they face.

The number one cause of job-related injury today—physical accidents—has little to do with genetic prophecy. But exposure to toxic chemicals, which ranks a close second, does. The industrial revolution and advances in chemistry have created substances that no living organism has ever before encountered. According to the American Chemical Society, we have festooned our air, water, and land with over 3½ million different chemical substances that do not normally occur in nature; about 63,000 are in common use in the United States. No longer do we have to travel to distant planets to find alien environments; we have developed one for ourselves right here.

Not all those chemicals are dangerous. One study of several hundred common chemicals found that only 25—substances like benzene, arsenic compounds, asbestos, nickel, tars, and vinyl chloride—have actually been shown to cause cancer in humans. Other chemicals may trigger other ailments, but most are harmless; we splash, spray, and smear dozens of them on our bodies daily. Only the few dangerous substances that do get through our defenses are the targets of prophecy.

Chemical Culprits

Nobody handles concentrated sulfuric acid carelessly. The consequences are immediate and absolute: If you spill it on bare portions of your anatomy, you will get burned, no matter what your genetic background is.

The toughest dilemma in the workplace is not with the kind of chemical from which every worker must be protected. The problems arise with chemicals that may be subtly toxic, that may do delayed and unpredictable damage to susceptible workers. For when a chemical is not obviously harmful, it may take years to pin down what it does and years more to identify the workers who are most at risk.

In 1895, the German physician Ludwig Rehn noticed that workers in the chemical dye industry had much higher rates of bladder cancer than people in the general population. He suspected that the disease was caused by the industry's use of massive amounts of arylamines—chemicals that are still vital to the preparation of rubber, plastics, textile and hair dyes, and other pigments.

Some 40 years later, scientists proved that a well-known arylamine caused bladder cancer in dogs. And 15 years after that, studies of the British dye industry confirmed the role of arylamines in human cancer. It took over 50 years for the chain of events to link suspicion and proof.

But implicating arylamines did little good. Not every dye worker develops cancer, even under similar working conditions. At first, researchers believed that the differences might be due to dietary habits. But soon other pieces of the puzzle began to fall into place. Animal experiments clearly showed that arylamines were not dangerous in themselves; they had to be transformed by the body into carcinogens. Dogs, which can carry out the transformation, are exceptionally susceptible to arylamine-induced cancer. Guinea pigs, which cannot, are resistant.

Because humans have the ability to transform arylamines, we are at risk for bladder cancer. But we are not all susceptible; some of us manufacture large amounts of an enzyme called N-acetyltransferase, or NAT, which attaches a bit of chemical to arylamines and de-activates them. People with high levels of NAT are called "rapid acetylators"; people with low levels are "slow acetylators." In the Caucasian population of North America, the distribution is almost exactly 50:50. Slow acetylators among other populations range from about 10 percent among Orientals to about 70 percent among Israelis.

Workers with low levels of NAT, that is, slow acetylators, are at greater risk of bladder cancer. But can the marker be used more widely? G. M. Lower of the University of Wisconsin Center for Health Sciences thinks so. He went to Denmark, where the ratio of rapid and slow acetylators is about the same as that in the United States, and examined bladder cancer patients in the urban population of Copenhagen. He found that slow acetylators accounted for twice as many victims as rapid acetylators, meaning that, in cities, slow acetylators are twice as likely to contract bladder cancer as rapid ones.

So far, Lower has not been able to pin down the reason for his findings. He theorizes that arylamines appear more frequently in urban environments, just as they do in the plastics, dye, and rubber industries. His hypothesis is supported by the fact that no differences in bladder cancer exist between fast and slow acetylators in rural areas, where arylamines are very rarely found.

Breathing Can Be Hazardous to Your Health

The respiratory system has a difficult job. Literally thousands of different substances pass through the lungs every day, and many are potentially dangerous. Today, researchers are trying to identify those harmful particles and to find out who is at risk.

Several markers used in predicting respiratory disease in the workplace are the same as those used for illnesses related to smoking. That should not come as too much of a surprise; cigarette smoke contains several hundred different chemicals and irritating bits of matter. A cigarette is practically a miniature chemical factory by itself.

AHH, the marker for lung cancer, is only one of many that may turn out to be useful in industrial preventive medicine. Perhaps equally important is a protein known as alpha-1-antitrypsin, or AAT; the disease it helps predict is chronic obstructive pulmonary disease (COPD), a combination of emphysema and chronic bronchitis.

COPD can be caused by infectious agents; but it can also be triggered by such irritating chemicals as cigarette smoke, air pollutants, and dust. One theory holds that irritation from these substances forces the body to release an enzyme called elastase. Elastase is designed to break down the irritants so that they can be excreted. But it also gnaws away at the walls of the lungs' tiny air sacs, where oxygen is transferred into the blood. As the walls break down, the lungs lose the capacity to transfer oxygen. Breathing becomes labored; and the symptoms of emphysema and chronic bronchitis appear.

Normally, people produce a second substance, AAT, which inhibits the action of elastase so that it leaves lung tissue alone. People with enough AAT tend to be more resistant to COPD; those without it tend to be more susceptible.

Two normal genes—called M genes—are needed to produce enough AAT to protect lung tissue against elastase. Unless lung irritation is excessive, people with two M genes are relatively safe from COPD. But a small percentage of the population carries at least one abnormal gene, called Z or S, instead of the normal M genes. Some ethnic groups are more at risk than others; among the Irish, about 9 percent carry at least one Z gene, while the percentage approaches zero among Italians and American Indians. If someone carries two abnormal Z genes, he is left with practically no protective inhibitor, and the elastase he produces simply starts chewing up lung tissue. Approximately 70 percent of ZZ-type people eventually develop COPD. Smokers with ZZ genes develop emphysema an average of nine years earlier than nonsmokers with the genes, and ZZ individuals are generally more susceptible than others to a whole range of respiratory irritants found in industry.

There is no question that those who carry a double dose of the Z or S gene are at increased risk. But what of the person who carries one Z and one normal M gene? Some studies have indicated that MZ people are not as susceptible as ZZs, but are more susceptible than MMs. Studies in West Germany of two groups of COPD patients come up with more MZ types than could be accounted for by chance. Other studies have provided conflicting data. Nevertheless, on balance, clinical investigations tend to show that MZ individuals who are exposed to smoke are more likely to develop respiratory complications than their MM co-workers.

Other AAT-related susceptibilities have been explored. The findings are still tentative because the number of patients studied is small, but the trends are the same as in the COPD study. Studies in cotton mills in South Carolina have indicated that workers who carry the ZZ or MZ genes are more prone to a respiratory ailment caused by inhaling cotton dust. And in Greece, low levels of AAT have been found in a larger number of tuberculosis patients than can be explained by coincidence.

Airborne respiratory irritants are among the oldest causes of job-related diseases. As far back as 1713, the physician Ramazzini noted that workers exposed to "vegetable dust" sometimes came down with a pneumonialike disorder. In 1932, more than 200 years later, the cause of a similar disease was traced to moldy hay. Since then, a variety of pneumonialike diseases have been discovered. The names may all be different, but the symptoms are much the same:

- Suberosis in wine makers from moldy cork
- Cheese worker's lung from moldy cheese
- Sequoiosis in lumber mill workers from moldy redwood dust
- Maple bark stripper's disease from moldy maple bark (in maple syrup collectors)
- Baggassosis in sugar cane workers from moldy sugar cane
- Wheat weevil disease in grainery workers and bakers from infested wheat flour
- Bird breeder's lung from pigeon and parakeet droppings

Research on the specific effects of the MZ gene combination on respiratory illness is still incomplete. Nevertheless, Hugh Evans, the Director of Pediatrics at the Jewish Hospital and Medical Center in Brooklyn, New York and a specialist in AAT disorders has said: "While I would never deny anybody with one Z gene a job, I would definitely advise that person to avoid a profession where he or she would come into contact with any respiratory irritants. Personally, if I had a single Z gene, I wouldn't work as a copper smelter or in a cotton mill; I'd learn other skills until the verdict was handed down."

In addition to the AAT marker, researchers are investigating the possibility that workers who are most at risk for respiratory illnesses have strong allergic reactions to airborne irritants. Quite possibly, their immune systems are tuned to marshal a powerful response to the invasion of particles; the pneumonialike symptoms that occur are, in fact, directly related to the strength of the immune system's response. So far, testing for these allergic reactions is useful only *after* exposure to the irritants. While the test may not be as helpful as the advance notice offered by a genetic marker, it does warn that breathing difficulties will develop if exposure continues.

Blood Feud

Between March 1968 and February 1969, some 4,000 young black men underwent Army basic training at an altitude of 4,060 feet in Denver, Colorado. On the first day of his training, a 21-year-old complained of faintness after a 20-yard low crawl. He lost consciousness and was dead on arrival at the local medical clinic. Another collapsed and lost consciousness after a 40-yard crawl and a 300-meter run. One hour after he regained consciousness in the hospital he fell into a coma; 24 hours later he was dead. A 19-year-old dropped after running a mile on the 21st day of training; a day later, he too, died. And still another arrived late for training, ran once around the barracks, and fainted. Eight hours later, he was the fourth to die.

Autopsies later revealed that all four men had swollen

blood vessels packed with red blood cells. The cells had a characteristic sickle shape.

These men did not have classic sickle-cell anemia; they were merely carriers of a single sickle-cell gene. The distinction is important. Sickle-cell disease occurs in individuals who have inherited *two* genes for an abnormal hemoglobin molecule called HbS. When a red blood cell that contains HbS hemoglobin loses its oxygen, the cell changes shape. Its normal pliable form becomes long, curved, and rigid, like a crescent or "sickle" moon. Because of their shape, the "sickled" cells pile up on one another and obstruct the flow of blood to vital organs. The neighboring tissue, lacking oxygen, begins to die.

Sickle-cell anemia, in which both hemoglobin genes are affected, is a serious disease. It is the most common form of hereditary hemolytic anemia, found in 15 out of every 10,000 black American children. But one out of every 12 American blacks carries only one copy of the gene; and that is the disorder that has caused all the controversy.

Carriers of a single gene for HbS are said to have the sickle-cell trait. They have minimal clinical problems; neither their average life expectancy nor the frequency of hospitalization is significantly different from individuals with normal hemoglobin. Yet they are not completely spared. Although their red blood cells are less likely to undergo sickling than the blood cells of people carrying two HbS genes, the problem usually affects between 20 and 40 percent of their cells. They do show evidence of sickling.

That is the catch. On the average, in an unstressed life, there is no difference between carriers of a single sickle-cell gene and normal people. But certain environmental triggers (lack of oxygen, higher altitude, or dehydration after physical exercise) can cause the red blood cells of single-gene carriers to sickle. Some physicians have suggested that people with the HbS marker should be warned against sudden drops in oxygen, noting that such occupations as mine rescue work and high altitude flying might be dangerous.

Still, a blanket indictment of the sickle-cell trait would be a mistake. If the 4,000 blacks in that Army training program had the same ratio of carriers as the general population, about 330 of them were carriers and had no problem whatsoever. For carriers of the sickle-cell trait, the critical issue is not the trait itself but the degree to which it can affect their red blood cells. The more the cells sickle, the more likely they are to become ill. The same simple blood test that identifies carriers can now also disclose just how much sickling can take place. For the four black soldiers in Denver, that information might have meant the difference between life and death.

The possibility that certain environments might cause illness among those with sickle-cell anemia, as well as among some carriers led the U.S. Air Force Academy and some industrial concerns to institute sickle-cell screening programs. The results have been mixed. The Air Force Academy, which requires the screening, used to automatically disqualify an average of five black students a year before one of them sued in 1980, alleging discrimination. It was only a matter of weeks before the Academy was forced to back down and agree not to disqualify any otherwise eligible black applicant simply because of his status as a carrier of the sickle-cell gene.

The screening attempts of industry have been less controversial. The DuPont Company, for instance, was asked in 1972 by its own Black DuPont Employees Association to begin a screening program. The test is voluntary; blacks can refuse to take it without jeopardizing their chances for a job; and DuPont does not deny jobs to those found to be carriers. According to Charles Reinhardt, director of DuPont's Haskill Laboratory for Toxicology and Industrial Medicine, "Workers in whom we find these traits are offered placement in areas where they cannot come in contact with the hazardous chemicals."

Nevertheless, even DuPont has come under some fire. Some critics contend that the company should be cleaning up the work environment instead of testing the employees, and they point out that, while other ethnic groups are at risk for different diseases at DuPont, only blacks are singled out for a national screening program.

Sickle-cell anemia is not the only blood disease that turns up in the workplace. The same problem that affects the Sardinian population during fava bean-growing season—hemolytic anemia—can also occur in the presence of industrial chemicals.

Only one in a thousand Anglo-Saxons carries the G-6-PD marker, but it appears in more than 10 percent of Filipinos, American blacks, and Mediterranean Jews. When those with the deficiency inhale chemicals as common as lead, carbon tetrachloride, benzene, naphthalene (in moth balls), and cresol, their red blood cells begin to explode. Recently, three equally common culprits have been added to the list; ozone, chlorine, and copper.

High concentrations of ozone are unhealthy for most people, but those with G-6-PD deficiency are especially sensitive. Ozone, a form of oxygen, is abundant at high altitudes and in city smog. When people with G-6-PD deficiency are exposed to a level of only one-half part of ozone per million parts of air (an amount commonly found in such major cities as Los Angeles, Chicago, and London) for as little as three hours, their red blood cells begin to disintegrate. Ozone has a characteristic smell similar to electricity. But concentrations as small as those that are dangerous can only be measured by instruments.

Chlorine is used to disinfect urban water supplies. But it can be poisonous to the susceptible individual. When chlorine dioxide is bubbled through drinking water, it is transformed into chlorite, which can cause hemolytic reactions. (Chlorite can be detected at the filtration plant.) At least one country has already responded to this health hazard; Norway now recommends that drinking water be filtered for chlorite during treatment.

Copper finds its way into the water supply when copper plumbing is used in areas where the drinking water is either too acidic or too alkaline. Copper's ability to trigger acute hemolytic anemia has pushed the National Academy of Science to suggest that people with the G-6-PD marker are probably susceptible to the metal and should avoid copper plumbing if they can.

The compounds that trigger hemolytic anemia all have one thing in common: They disrupt the normal functioning of the red blood cells. In large concentrations, they can seriously disturb even normal blood cells, but normal cells, at least, usually can adapt to average stress.

Who is Responsible?

The discovery of better and more accurate markers foreshadows the speedy expansion of industrial screening. Soon we may find ourselves caught in an ever-tightening spiral: finding markers for more subtle problems, discovering larger numbers of susceptible workers, labeling more and different types of environments as hazardous. It is not inconceivable that, through screening, industry will become the modern counterpart of Diogenes as it searches for a perfect worker.

But perfection in industry is a relative trait. Is the perfect worker someone who can function safely in nearly any environment? Or is it someone who can remain healthy only after companies make a major effort to render the workplace safe? Who is responsible for safety: the worker or the company?

That critical question has given medicine extraordinary political leverage in industry, but it has also permitted politics to distort the traditional physician-patient relationship. The same information that may help an employee decide to avoid a toxic substance may also be used by an employer to force a worker out of a job.

The issue is job discrimination, or what has been euphemistically labeled "protective exclusion." So far, only one "genetic marker," sex, has been used extensively for this purpose; women have traditionally and consistently been excluded from jobs requiring excessive physical exertion. And more recently, at least a dozen major corporations, including Dow Chemical, General Motors, Monsanto, and Firestone Tire and Rubber, have excluded fertile women of childbearing age from certain jobs in order to protect the potential fetuses from harm.

This policy has now been sharply attacked. In October 1978, four women who had undergone voluntary sterilizations in order to keep their jobs at American Cyanamid's lead pigment plant in Willow Island, West Virginia, sued the company. The women charged that by forcing them either to undergo the surgical procedure or to lose their jobs, American Cyanamid had violated their civil rights.

American Cyanamid and other chemical companies pointed out that unless they protected the potential fetuses, they would be vulnerable to claims for damage if babies with defects were born. The women contended that fetuses could be harmed through the male as well as through the female, and that men were not being restricted in any way. Furthermore, the women themselves were in no danger; the possible hazards cited by the company extended only to their potential children. And some had no intention of getting pregnant.

The case went to trial and American Cyanamid lost the first round. But no court case can answer all the questions this suit raises; no set of Federal regulations will satisfy both sides. Chemical companies simply cannot protect all their workers, and their workers' future offspring, from the possibility of harm. In the particular case of American Cyanamid, the company's liability in the suit was caused neither by willful neglect nor by carelessness or indifference, but by its interference in the lives of a group of people so that it might protect the lives of others, even when they had not yet been conceived.

The American Cyanamid case raises one of the most difficult problems faced by industry today: When only *some* people are susceptible to an environmental hazard, what should be the standard of effective protection? Should a company provide a work environment that is completely safe for the "average" worker? Or must it extend its efforts to protect even those who are most susceptible?

The same dilemma can arise for any genetic trait that makes a worker susceptible to chemicals in the environment. The most common job-related problem, for instance, is a group of skin disorders that is lumped together under the catch-all title, "industrial dermatitis." Of every ten claims paid by insurance companies for industrial disease, seven are for some variety of dermatitis. Most stem from contact with substances that are safe for most workers. Two easily observed genetic markers serve as warning signals: race and oily (as opposed to dry) skin.

Light-skinned people, especially those of Celtic origin, are more susceptible to skin irritants than dark-skinned people. The difference has nothing to do with whether or not blacks have "thicker skin"; investigators have actually filed the outer layers off and found that blacks remain more resistant.

People with oily skin are more sensitive to certain oils than their dry-skinned co-workers, but they are less sensitive to industrial solvents like alcohol and turpentine.

Some companies look for these markers. DuPont in particular advises workers with oily skin to be more careful in working with industrial oils, to wash more often, and to use protective clothing.

But is that enough? Since most workers have no problem with oils and solvents, is the company justified in assuming that the process of protection can be left up to the workers themselves? Or should it be taking steps to ensure that these substances can never come in contact with sensitive skin?

The companies have argued that there are limits to what they can do. Some problems are still technically insoluble and the cost of alleviating others would be prohibitive. At the height of the controversy over American Cyanamid's female workers, the editors of *Chemical Week* wrote:

> It makes no economic sense to spend millions of dollars to tighten up a process that is dangerous only for a tiny fraction of employees—if the susceptible individuals can be identified and isolated from it.

The other side emphasizes the responsibility of the employer to those whose lives he has bought for eight hours a day, as well as to those whom he cannot afford to hire because the environment he has provided could cause them (and, by legal extension, him) harm. It also warns that screening programs may give companies the opportunity to ignore their duty to clean up the workplace as they emphasize deficiencies in the workers. Companies may assume that it is simpler and cheaper to rid the environment of the susceptible worker than of the dangerous substance.

Nevertheless, the nature of the workplace makes it one of the best locations for screening programs to take place. Studies have found that more than 90 percent of all workers in a factory participate in voluntary screening programs at work, as opposed to about 30 percent in identical programs in the community. A program conducted on the job requires no lost travel time. It makes counseling, observation, and medical follow-up much simpler. And the companies often provide staff doctors and nurses who are aware of their particular environmental problems to care for those identified as susceptible. Because the American labor force includes about 45 percent of the population, carefully organized industrial programs can reach most of those at risk.

Screening in industry ultimately protects both management and workers. Pinpointing who is at risk and cleaning up the environment are only beneficial when both sides understand the size and extent of the danger. True protection for everyone concerned stems from a sharing of responsibility: a full effort by industry to clean itself up where it can; and a recognition by workers that their health and safety depends to a great extent on the choices they make.

6

DIETS AND DRUGS

What is food to one man may be sharp
poison to another.

LUCRETIUS

Whereas substances in the workplace enter our bodies primarily through the skin and lungs, products from the supermarket reach us through our stomachs. Eating, in a sense, is just another way that the environment gains access to our genes.

The relationship between food and health has long been obvious. We have known since the nineteenth century that a lack of vitamin C triggers scurvy, since the early twentieth century that a deficiency of vitamin D causes rickets; and television has now given us the chance to witness the horror of children suffering from malnutrition. From the moment green leafy vegetables were first forced between our lips, we have been privy to the litanies of food and health: Spinach is good for you; candy is bad.

Now those simple rules are growing more complex. We talk of low cholesterol and high cholesterol diets, of meals stocked with roughage and bran, of meatless days and carrot juice, of multivitamin supplements to replace the nutrients that modern food processing techniques squeeze out of what we eat. Proselytizers choke the airwaves with advice on how we can eat better to live longer. Entire magazines are dedicated to the proposition that one diet plan or another can offer us the ideal bodies we so desperately want. Literally thousands of books beckon to us from bookstore shelves, urging us to balance or alter our eating and living habits to take into account the latest in a long line of nutritional theories.

But they all fall short on one important level: They do not discriminate. They are invariably aimed at the "average individual," a characterization that in itself makes no sense, since no individual is truly average. They provide blanket advice designed to help everyone, not recognizing the variations that exist among people—variations that alter their needs and responses to different foods.

Like everyone else, geneticists used to view food as a general means to a general end: dinner. But in the past decade, they have begun to examine our eating habits and our tendencies to gain weight with a critical eye. They are discovering how foods can be harmful, not necessarily to everybody, but to specific groups who may be susceptible.

Blame It on Your Genes

Our society is obsessed with the lean look. Our models are ideally slim; obese people comprise perhaps the most blatantly stigmatized group in society (after all, the logic goes, they could do something about their problem, if they only had the will power); millions of bottles of diet soda and tons of cottage cheese are consumed annually to make sure that we can fit into our designer jeans.

Yet for some reason, we, as a society, do not lose weight; the average individual is actually fatter than ever before. Our failure has been blamed on the nutritional problems of dieting, the psychological aspects of obesity, the inability of those caught in the weight spiral to realize that losing weight is not painless, but requires discipline, exercise, constant ego reinforcement, and a dedication to keeping the weight off once it has been lost. This is a far cry from the claim of some manufacturers of commercial products that taking a certain pill or wearing something resembling a diver's wet suit can make all the difference in the world.

A recent survey found that Americans are among the most overweight people in the world. Only the citizens of Rome,

Italy, come close to us in the amount of extra fat their bodies carry. There is little doubt that much of this obesity is related to the higher standard of living we enjoy—more cars, more leisure, more calories—and that the amount of food we eat is directly linked to our excess weight. But are these the only factors?

In the early 1970s, many scientists believed they were. They found that the number of fat cells we have depends in large part on the amount of food we consume as children. If we are too well fed, we develop too many cells. Then, as we grow, the cells remain. Their number can be increased by overeating. Dieting causes them to shrink, but not to disappear. This is how too much to eat early in life may almost inevitably lead to some adult obesity.

The role of the fat cells in obesity is now generally accepted. But it is slowly being buttressed by questions surrounding the contribution of the genes. Genes *are* involved in helping to determine whether or not you can stay slim. Everyone is familiar with cases of grossly overweight people who suffer from hormonal imbalances. The fat lady in the circus is one example; the 1,069-pound man in the *Guinness Book of World Records* is another. But they are rarities. Nobody decides to eat less for fear of ending up looking like them.

It turns out that all these factors are important. All contribute to the general state of obesity that exists in the United States.

When scientists first suspected that obesity could have a genetic component, they found themselves in a corner. Most studies pointed to dozens of environmental factors, all of which made it difficult to isolate the genes and experiment with them. Nevertheless, the scientists realized that even people of normal weight put on pounds at different rates when they consumed identical amounts and kinds of calories. They were struck by the implication that genes could be deeply involved in weight gain. As usual, when confronted by a difficult genetic problem, they turned to a mouse for assistance. In this case, they happened to find a fat mouse.

The obese mouse carries two copies of a gene called the obese, or *ob*, gene. Littermates who are identical in every other way but lack the *ob* gene remain lean. When scientists compared the two types of mice, they found that, if given the same amount of calories as the lean group, the obese mice retained a greater portion of the energy they took in. In short, the lean mice were better equipped genetically to burn off the energy they didn't need.

Soon the genetic defect caused by the *ob* gene was pinpointed. The problem was in sodium potassium ATPase—an enzyme also called the sodium pump because it regulates the amount of sodium and potassium inside cells by pumping the salts through the cell walls. The pumping process is critical for maintaining proper cell function throughout the body; it also consumes large amounts of energy. In fact, some scientists have calculated that between 20 and 50 percent of a cell's heat is generated by the activity of this pump.

Could the defective enzyme be a marker for obesity? Could humans have the same defect as mice? Mario de Luise and his colleagues at Boston's Beth Israel Hospital undertook a study of obese patients to find out. They examined 23 patients, ranging in age from 22 to 49 years old. The seven men and sixteen women were two and one half to five times heavier than the weights most nutritionists would consider ideal for their heights and body types.

Routine laboratory tests all proved normal, except for slightly elevated sugar levels in two of the patients. There were no clinical indications of hormonal imbalances. And so de Luise and his team checked the activity of the sodium pump. Two of the subjects' problems turned out to be unrelated to the activity of the sodium pump; they are still being studied. But the 21 others had lower pump levels than all but the lowest of a normal group of controls.

One question remained—the chicken-or-the-egg problem that often confronts scientists who find two things that relate to each other: namely, did the people have less active pumps because they were obese; or did they become obese because they had less active pumps? Only the second possibility held the promise that the level of activity of the sodium pump could be used as a genetic marker.

The solution was simple; the scientists examined the pump's activity after the patients had lost weight. About 18 months after the patients began a strict diet, they had succeeded in losing an average of 18 pounds for every hundred they had originally weighed. But there was *no* change in the pump's activity; the difference in weight had no effect on the amount of energy the pump consumed.

The Beth Israel work alone is not enough to confirm the

usefulness of the sodium pump as a marker for obesity; new studies must be undertaken to answer questions that are still unanswered. But animal studies of the pump's activities from birth support de Luise's findings. Together, they strongly suggest that people can have either active or sluggish sodium pumps. The type of pump we have can tell us whether we will have difficulty in burning off excess calories—in other words, whether we are prone to at least one type of obesity. In practical terms, it means that certain genetically-based problems of obesity require dietetic treatment with an eye to the genes; visits to psychiatrists and self-help groups may improve a person's willingness to hold to a strict diet but will have no effect on the underlying causes of the problem. In philosophical terms, it means that blanket statements about obese people are grossly unfair. They do not necessarily lack will-power or motivation. They do not necessarily develop a lessened concern for their appearances. Their bodies merely react differently to an environmental stimulus—food.

While some people worry about how much weight they gain, others are concerned about their *inability* to put on a few pounds. Some suffer physical problems when they try to eat more food. Those with less severe problems may have only mild discomfort and diarrhea. Others, suffering in the extreme, may have gross disturbances in structure and function: weight loss, muscle wasting, and skeletal disorders. Despite the wide variety of clinical signs, many of these people suffer from the same syndrome, celiac disease, caused by the same genetic trait, the inability to absorb food from their intestines.

The reasons for celiac disease are still far from clear. But at least one environmental trigger has been identified, a protein called gluten. Many who suffer from celiac disease cannot tolerate foods made from wheat, rye, and other types of flour (bread, cakes, cookies, pasta), as well as beer and ale, which also contain gluten. When these foods are eliminated from the diet, appetite improves, diarrhea disappears, and weight begins to increase.

Gluten causes problems when it cannot be fully digested and excreted. It then appears to have some toxic effect on the cells of the intestinal wall. After a while the wall itself begins to undergo a physical change, becoming thicker and less capable of absorbing nutrients. At that point, the disease begins to have clinical implications.

Just how many people suffer from gluten sensitivity is not known; about 70 percent of the reported cases are women. But because the symptoms can be so mild, many people never seek medical help; and even those who do are often misdiagnosed. People can go through life with minor intestinal discomfort, never realizing that the problems are caused by the wheat proteins they consume every day.

Chronic progressive problems like celiac disease are natural targets for genetic prophecy. Cutting back on one's consumption of wheat may be an inconvenience, but if it prevents or cures a potentially serious illness, it is worth the sacrifice.

So far, several HLA markers have been found in people who are sensitive to wheat. Although they are probably not *directly* related to the problem, their association can be quite powerful. One of the HLA markers—HLA-Dw3—appears in 98 percent of those who have the sensitivity. Someone with the marker who habitually eats wheat has 278 times greater chance of getting celiac disease than someone without it. But if that same person begins to avoid wheat and its products early in life, the risk drops to normal levels. Others with the marker may continue to enjoy wheat and grain products until the early stages of the disease are disclosed by regular checkups that pay particular attention to the possibility of intestinal damage. For those with the sensitivity, food companies have developed flour that is gluten-free.

Modern technology has also produced a substitute for another product to which millions of people are sensitive: milk. Lactose, or milk sugar, is an irritant to those who are genetically incapable of digesting it. Because they lack a critical enzyme, their systems cannot tolerate milk. They suffer abdominal cramps, bloating or distension, and diarrhea.

While only about 5 percent of the adult white population in the United States is sensitive to milk, as many as 60 to 90 percent of American blacks and a large number of Northern Europeans and Orientals have the deficiency. The high incidence of the disorder among American Indians explains why they used to paint their adobes with the powdered milk they were given by the government. They weren't ungrateful; it just made them sick.

In many of these people, symptoms of milk intolerance do not occur until puberty or late adolescence. Although a simple blood test can locate a genetic marker that pinpoints this predisposition, physicians have found little reason to bother with the test. Milk intolerance is not life-threatening; it does not progress irreversibly. Simply avoiding milk and milk products, or using the milk substitute (in which the lactose has been broken down) is enough to clear up the symptoms once they begin.

Ulcers

Twenty million Americans have duodenal ulcers. In 1979 alone, they spent $2.2 billion on health care. But the real costs, in pain and suffering, far exceed those numbers.

For as yet unknown reasons, most people who are susceptible to ulcers produce too much stomach acid. The acid that is not used to help digest food starts to work on the walls of the stomach and intestine. Normally, the walls are protected by a secretion that neutralizes the corrosive effects of the acid; but if too much acid is produced, the defense falters. If the acid burns deeply enough, it can reach the blood vessels and cause intestinal bleeding, which is a clear sign of serious digestive problems.

Duodenal ulcers were one of the first disorders to have a genetic marker mainly because the marker, ABO blood type, was discovered and studied so early. The major ABO types have been known since 1900. Testing for them has always been so simple that we have gathered comprehensive statistics on all sorts of ethnic, racial, and geographical groups. We know, for instance, that 44 percent of white Americans have blood type O, compared to only 25 percent of Pakistanis and over 85 percent of Chippewa Indians. We also know that blood type O points to a slight but significant predisposition to duodenal ulcers; those who have it are 1.4 times as likely to get ulcers than those with types A, B, or AB.

The correlation of blood type O to ulcers is low; it is probably not directly linked to the mechanisms of acid production that cause the disorder. Recently, however, researchers have begun to discover exactly what those mechanisms are. In the process, they have uncovered a marker that may be extremely useful for prediction.

In 1971, Jerome Rotter of the Harbor-UCLA Medical Center carried out a collaborative study with colleagues from the University of Liverpool, England, combing the records of patients at the Broadgreen Hospital in Liverpool for a list of patients with duodenal ulcers. They found 123 people whose families agreed to participate in a broad study in which all brothers and sisters would be medically examined.

The investigators examined the family members for both ulcers and high levels of pepsinogen 1, an enzyme produced in the stomach. They identified individuals with ulcers in over half the families. And the incidence of ulcers in these families was extraordinarily high; of the 111 brothers and sisters in families with a history of ulcers, 29 (more than 25 percent) had the problem. Obviously, just being a member of the family somehow increased the risk.

But the real payoff came when those same 111 family members were checked for their pepsinogen 1 levels. Half had high levels, half had normal levels; but almost all the ulcer victims had high levels. Of the 29 individuals who were found to have ulcers, 22 produced too much pepsinogen 1. And even among the seven who did not, three fell on the high side of the normal range. A full 40 percent (22 out of 55) of those with high pepsinogen levels developed ulcers, compared to only 12 percent (7 out of 56) of those with normal levels.

When we can predict that individuals in certain families have two chances in five of contracting duodenal ulcers, we are beginning to reap the benefits of genetic prophecy. Pepsinogen 1 levels can be measured in the blood. Members of families with a history of duodenal ulcers who produce too much of the enzyme may be encouraged to reduce their chances of getting ulcers by avoiding foods like coffee and alcohol (which are known to increase acid secretion in the stomach) and by staying away from the light midnight snacks that encourage the stomach to produce digestive juices without giving it enough material to digest.

No studies have yet been performed that evaluate the importance of a high level of pepsinogen 1 and blood type O together as markers for ulcers. These and other studies may ultimately pinpoint a smaller group which carries an even higher risk.

Abnormal levels of pepsinogen 1 and specific blood groups have a relationship that extends beyond ulcers, however. Recently, they have turned up as markers with important implications for another disease: stomach cancer.

Three decades have passed since a significant association was reported between stomach cancer and blood type A. But the risk for people with blood type A was only slightly higher than that of the general population and the usefulness of the blood type as a marker by itself was questionable. Researchers began to look for other ways to identify a smaller population with a higher susceptibility.

In 1980, they found one. Abraham Nomura of Honolulu, Hawaii took blood samples from 7,498 Japanese men in the late 1960s. Ten years later, he tested them for stomach cancer and found 48 cases. Among the markers examined was the enzyme which is increased in ulcer patients, pepsinogen 1. One-third of the cancer victims had *low* levels of pepsinogen 1, compared to about one-sixteenth of the normal population. And all those with low levels had the same type of cancer, known as "intestinal mixed-other." Nomura calculated that people who are under 60 years old and have low levels of pepsinogen 1 carry a risk of stomach cancer that is more than 20 times higher than that of the general population. Although little is known about harmful dietary factors, successful treatment of stomach cancer depends partly on catching it early enough. The discovery of the pepsinogen 1 marker should at least give those at risk a better chance of being diagnosed early.

Toward a Personalized Diet

Just as no job is suitable for everyone, no diet can serve everybody's nutritional needs. Genetic markers can act as early warning signals for our personal eating habits. In particular, they may be useful for uncovering conditions that might not necessarily drive us to seek medical attention. A small percentage of older women of Scandinavian, Irish, and English descent, for instance, suffer from a blood disorder called pernicious anemia. They tire easily and develop shortness of breath, frequent headaches, and slight palpitations, the kinds of symptoms that are all too often dismissed by friends and physicians as stemming from some minor neurosis.

But pernicious anemia, while rare, is an identifiable genetic disorder. It arises when the red blood cells lack a specific protein that allows them to absorb vitamin B_{12}. The cells need the vitamin; without it, they become bloated and heavy. Ultimately, their inefficiency can disrupt the function of peripheral nerves; and pernicious anemia can cause severe neurologic problems.

The disorder strikes only about one out of every 200 older women of Scandinavian, Irish and English descent. The only genetic marker thus far discovered, HLA-B7, merely doubles the odds to one in a hundred—not nearly high enough to concern either physician or potential patient. Nevertheless, its tendency to run in families and the presence of a defective protein makes it likely that investigators will soon discover a causative genetic marker (perhaps the protein itself) that can pinpoint those at risk. At that point, prevention may be simple; the ingestion of a larger dose of vitamin B_{12} to offset the body's inability to absorb it.

Pernicious anemia is just one of a host of disorders that links foods and nutrients to genes. In the last decade, similar connections have focused attention on some uncommon genetic reactions to some very common foods:

- Chocolates, which contain a chemical called phenylethylamine, can cause migraine headaches in those who have low levels of monoamine oxidase, an enzyme that can transform phenylethylamine to a usable, digestible form.

- Nearly all cheeses contain the chemical tyramine, which can trigger migraines in people who cannot transform it into a form the body can excrete. The enzyme that causes the transformation, tyramine transaminase, can be measured in the blood. The lack of the enxyme predisposes one to migraines.

- Iron-fortified foods (like some white breads) can cause an iron-storage disease called bronze diabetes in those with defective iron-carrying proteins in their blood. HLA-A3 identifies people who are nine times more likely to be affected than the general population.

- Vegetables like cabbage and Brussels sprouts contain substances (such as phenylthiocarbamide, or PTC) that may cause goiters, apparently by interfering with the thyroid's ability to gather iodine. These substances taste strongly bitter to people who carry a dominant "taste" gene. Because nontasters crop up more frequently than expected among patients with non-toxic goiters, some scientists have theorized that nontasters may eat more goiter-inducing foods because they cannot detect their bitterness. Others suggest that tasters have a built-in early warning system; they may be more susceptible to the goiter-causing compounds—but shy away from them because of their unpleasant taste. The connection between thyroid abnormalities and nontasting is also found in disorders other than goiters. (Testing for the ability to taste PTC is a simple matter of dabbing some on the tongue and deciding whether the sensation is unpleasant.)

Genes and Drugs

King George III of England, who ruled the Empire during the American Revolution, was prone to periodic fits of madness. His symptoms were such that psychiatrists today might term him a paranoid schizophrenic; they were so severe that, during one particularly long bout, Parliament debated whether the King should be declared permanently ill and replaced.

The pattern of his illness and the genetic heritage he passed on have convinced medical historians that George III was not suffering a form of mental illness. Rather, he was predisposed to porphyria, a disease that can be triggered by such substances as alcohol (as was probably the case for him), barbiturates, and sleeping tablets. Porphyria is also easy to recognize clinically; people suffering from it pass wine-colored urine.

Porphyria is a genetic disease that appears with greater frequency than usual among the white population of South Africa. That country has accepted its burden and has passed a law that requires physicians administering barbiturates to test their patients first for susceptibility to porphyria. This marked the first time that testing for a genetic sensitivity to drugs has been legislated into national practice.

In a sense, drugs are simply a more potent form of food. They are both used to maintain or restore the balance of health; and they have common origins: some plants were ingested because they tasted good or provided nutrition; others were chewed on to counteract disease. Even today, 80 percent of all our pharmaceutical products are extracted from plants. Their chemicals are simply more concentrated, purer, and more powerful than those in food.

Drugs are prescribed to prevent or cure disease, but they can also cause it. Everyone knows of deaths caused by an overdose of sleeping pills or an allergic reaction to penicillin. But most people do not realize that something can go wrong nearly every time a drug is prescribed.

Physicians recognize the dangers. They even have a word—iatrogenic—that describes adverse conditions that occur as the result of medical treatment. They have found that hospitalized patients, for example, receive an average of 10 different drugs during their stays, and that approximately 7 percent suffer undesirable reactions. In addition, as many as 20 percent of all patients suffer adverse drug effects at some time during their lives. When a drug causes an iatrogenic disease, physicians have learned to ask why.

The search for answers begins every time a new drug is discovered. It costs a company about $60 million to bring a single new drug to the market. Before it hits the local pharmacist's counter, the drug is examined for its chemical properties (how long it is stable, how fast it dissolves, at what temperature it decomposes), as well as for its biological properties (how rapidly the body absorbs it, which cells it affects, how long it stays in the tissues). By tagging it with radioactive labels, scientists can follow its path from the moment it enters the body to the moment it leaves.

Nevertheless, these laboratory tests describe an *average* for the distribution of a drug in the body and the rate of its elimination. In patients, these properties are influenced by various factors: age, weight, diet, and other drugs. We are warned not to mix alcohol and sleeping pills, and we cut a dose of aspirin in half before giving it to young children.

Genes, too, can dictate different reactions to the drugs we ingest. Studies of identical twins, for instance, have consistently shown that the way we handle drugs depends to a great extent on our heritage. Different people can take from a day to a week to clear an equal dose of the same drug from their bodies. But identical twins' rates of elimination are almost exactly the same. Because of the vast array of factors that can change our response to drugs, physicians are now trained and encouraged to devise drug regimens that are tailored to the particular patient. And genetic markers are helping them in that task.

Sometimes those markers can be as straightforward as the ABO blood types. The link between oral contraceptives, blood type A, and blood clots, for example, was discovered in the simplest of experiments.

In 1969, physicians who were monitoring the blood types of patients admitted to the wards of three Boston hospitals noticed that fewer patients with blood type O received anti-coagulant drugs (used mainly to treat blood clots) than those with other blood types. This tiny lead and the controversies surrounding oral contraception prompted them to arrange a cooperative study among research groups in the United States, Great Britain, and Sweden to determine whether their original discovery meant anything.

It did. When the three surveys had been completed, they all showed that young women with blood type A were slightly more prone to clots than those with blood type O. But that slight increase in risk actually doubled if the A-type women also took birth control pills. In all, women with blood type A who also took oral contraceptives were two and one half to five times as likely to develop blood clots as women on the pill with blood type O.

The blood type A marker for predisposition to clotting differs from most other markers for drug reactions primarily because the illness it can cause stems from constant use of birth control pills. Other drugs often trigger more immediate, if not more tragic, results. Most of them are foreign compounds that the body has never before encountered; we cannot always know ahead of time how we, as individuals, will react. Despite the extensive tests that a pharmaceutical company may put a drug through, it is still too easy to overlook the rare individual who may have a bad reaction.

The muscle relaxant succinylcholine typifies the rare yet dangerous genetic susceptibilities that may occur. Patients undergoing surgery must often be completely relaxed before they can undergo a variety of surgical procedures—the manipulation of an arm or a leg, perhaps, or the insertion of a tube down the throat.

Succinylcholine makes these procedures possible. It causes muscles to relax completely, but its effects usually last about two to four minutes, which is not long enough to cause problems. From its first application in the early 1950s, physicians praised its safety. By 1952, several thousand patients had received the drug without ill effects. Then the trouble began.

In 1952, two patients at St. Bartholomew's Hospital in London were given the standard dose of succinylcholine. Four minutes passed; the drug-induced effect of muscle relaxation (or paralysis) showed no signs of abating. In both cases, the surgeons went quickly to work. Prolonged paralysis would involve the respiratory system; if they failed to compensate for it by getting air to the patients' lungs, the two would die.

The paralysis lasted 20 minutes for those two particular patients. For others since, it has gone on for up to three hours. And while there are no reports in the literature of deaths from prolonged reaction to succinylcholine, most physicians acknowledge that some have probably occurred; the surgeons involved doubtlessly considered it in their best interests to keep the cause of death quiet.

Thus, a drug that seemed perfectly safe when it was given to one or even two or even a thousand patients was found to be dangerous in about one in every 2,500 cases. The susceptible individuals, missed by toxicity studies when the drug was originally tested, began to turn up.

We now know that susceptibility to succinylcholine is caused by a genetic variant. But while a simple blood test allows us to test for it if we wish to, the rarity of the variant makes that particular option economically unsound. Today, surgeons continue to give succinylcholine to all patients who require it during surgery. But they are constantly on the alert for the first sign of susceptibility, ready to apply artificial respiration if it is needed.

As we improve our methods of researching drug gene interactions, new correlations continue to crop up. One of the more recent involves the AHH system, the genetic marker for lung cancer in smokers. Hitoshi Shichi of the National Eye Institute has found that high levels of AHH in mice coincides with the formation of cataracts in those exposed to large doses of acetaminophen, the active ingredient in the over-the-counter pain suppressant, Tylenol. Although cataracts can be triggered by other drugs as well, Shichi discovered that his animals began to develop them within six hours after he had administered large doses of acetaminophen. It appears that the drug is rapidly converted into some other, as yet unknown, chemical in mice with high levels of AHH; it then flows through the bloodstream to the eye, where it interacts with the lens to produce the disorder.

A similar connection between acetaminophen and cataracts in humans has not yet been established, but it seems possible. Many people take the drug daily; and it may just be that the high incidence of cataracts in older people is partly due to the prolonged use of acetaminophen to counteract the pain of arthritis. Cataracts can be induced by other chemicals as well, from the naphthalene in moth balls to various substances floating around dye and chemical factories. The search for a genetic marker to pinpoint those who are susceptible is very recent; AHH may well turn out to be that marker.

Other drug-gene interactions are being discovered almost daily:

- The acetylator type—the same marker that pinpoints propensity for bladder cancer in workers in the plastics, rubber, and dye industries—also can affect responses to different drugs. Fast acetylators (people who break down the chemical arylamine easily) may develop liver poisoning from taking too much isoniazid, the drug of choice for tuberculosis; slow acetylators may risk damaging their peripheral nerves with the same drug.

  Slow acetylators also seem to be more prone to lupus erythematosus, a terrible disease in which the body's immune system turns against itself. In at least some cases, the environmental trigger has been hydralazine (also known as Apresolin), a drug used to counter hypertension.

- Aplastic anemia, a fatal disease of the blood, can be triggered by the antibiotic chloramphenicol. People whose bone marrow cells have greater difficulty in synthesizing DNA when they are tested in the presence of the drug are susceptible.

- Toxic reactions to sodium aurothiomalate (used in the treatment of rheumatoid arthritis) can be predicted on the basis of HLA type. Studies carried out at Guy's Hospital in London in 1980 showed that 19 out of 24 patients with toxic reactions carried HLA-B8 or Dw3, with a 32-fold risk over others.

- Warfarin and dicumarol are anticoagulants, used to

treat hundreds of thousands of people who have had pulmonary embolisms or deep vein thromboses. Some people are sensitive to the drugs; AHH, or a similar group of enzymes, may someday be used as a marker to determine who is at risk.

- Severe hemolytic anemia in those with G-6-PD deficiency can be triggered by everything from sulfa drugs and primaquine (used to treat malaria) to antipyrene (an analgesic used for swimmer's ear) and Furdantin (a bacteriocide).

By now it should be clear that anyone with a relative who has had a severe drug reaction should tread cautiously when around drugs. The genetic link to drugs means that reactions tend to run in families. Because many reactions are so rare, it is our responsibility to take notice of any that might apply to us. If your family history includes bad reactions to some drugs, it is essential that your physician be told. You may never confront your own drug sensitivities. But you might, and the use of genetic markers could save your life.

7

GOOD GENE, BAD GENE

Does anybody know what will be best for
mankind centuries or millennia hence? No heredity is
'good' regardless of the environment.

THEODOSIUS DOBZHANSKY

It was, from all accounts, a wonderful party. Among the
guests were Isadora Duncan, the dancer, and George Ber-
nard Shaw, the Irish playwright. Somehow they managed to
sit next to each other at dinner. Isadora Duncan seized control of the moment. She turned to Shaw: "Wouldn't it be
wonderful if we could have a child with my looks and your
brains?" she asked. Shaw considered the possibility. "Yes,"
he answered slowly. "But what if it had *my* looks and *your*
brains?"

Even before genetics became a full-fledged science, it constantly forced us to make value judgments. Parents-to-be
hoped that their children would inherit what they considered
their "best" traits, while praying that they would not be
cursed with less desirable family attributes. From the beginning, people formed clear ideas as to what constituted good
genes and bad.

In the extreme, the worst genes are easy to identify. A
gene that prevents an embryo from maturing into a live baby
is clearly deleterious. One out of five of all conceptions ends
in spontaneous miscarriage, and most geneticists agree that
genetic factors play an important role in this loss. The ratio is

108

probably even higher than that; defective genes are also implicated in the inability of some fertilized eggs to attach themselves to the uterine wall, which is a failure that we cannot yet detect.

General agreement also exists about the value of genes that may lead to gross physical abnormalities. Gene-directed characteristics like the absence of arms and legs may be compatible with survival, but only if society takes extraordinary steps to sustain the life of the affected child.

Incest

We all probably carry blueprints for characteristics that are incompatible with life. These genes have been termed our *genetic load*. While they may be lethal, they are recessive and do not express themselves as traits unless we happen to inherit two that are identical. Through the pain of experience, every society learns what happens when these genes are expressed. People have long realized that children born to parents who are genetically related die more often and earlier than other children. The closer the kinship between the parents, the worse the outcome.

A study of church records in the District Morbihan in France indicated the danger to children born of related parents. When parents were unrelated, the frequency of stillbirths and neonatal deaths was approximately 4 percent. Second cousin marriages almost doubled the number; and children born of marriages between first cousins died more than 11 percent of the time. The figures for deaths during early childhood were equally revealing; about 9 percent of the children with unrelated parents, about 11 percent of children of second cousins, and over 15 percent—one out of every six children—among first cousins, died.

Since earliest times, societies have responded to the knowledge that incest can lead not only to earlier death for offspring but also to severe disorders of both the mind and the body. That germane piece of genetic information is at the root of the nearly universal prohibition on incest.

Modern genetic theory can now explain what common sense has told us over the centuries. The development of a human embryo is like the creation of an automobile from two sets of blueprints. Directions for building each part may exist; but not all the directions lead to usable parts. Blueprint A, while reasonably complete, may have garbled instructions for making a windshield wiper—a problem solved by clear information in blueprint B. Meanwhile, blueprint B may give directions for a horn with misconnected wires and a useless shape, a component that can easily be built from directions in blueprint A. By building the components using only the worthwhile information—windshield wipers from blueprint B and the horn from blueprint A—we can construct a complete, running automobile, either discarding or not building the faulty parts in the process.

In each case, the faulty information never appears in the completed car; that information is hidden, or *recessive*. The information from which the parts are actually built is *dominant*. Clearly, if both blueprints A and B carry faulty information for the same part, the part would be unusable.

Human beings carry an average of four to eight recessive genes, any one of which may lead to problems if it is paired with an identical gene. When two parents pass on the same recessive gene to their child, the defect appears. Normally, the chances of two people, chosen at random, having the same defective gene is extremely small. Scientists estimate that we have anywhere from 50,000 to 100,000 pairs of genes, any one of which might be defective. It is only when individuals from the same genetic line mate with each other that the two defective genes are more likely to appear in the off-spring. That is the main reason for the taboo on incest.

While it is easy to come to an agreement about genes that kill, the vast majority of genes are less dramatic in their effects. They crop up in a number of subtle guises, many of which are not necessarily detrimental. Distinguishing good genes from bad can be an elusive exercise.

We can view genes as bad in three different ways:

- We can compare them to other, better genes.
- We can view them as both good and bad at the same time.
- We can judge their value according to the circumstances in which they are expressed.

Comparing Genes

Comparing one gene to another is a question of perception and personal values. It can be a simple matter of preferring one alternative over another when neither is clearly harmful or beneficial.

When genes are compared, their value may change, depending on the genes to which they are compared. If, for instance, someone is born with the ability to distinguish among three fine shades of red, while the average person can only discern two, we might assume that he has inherited the "good" genes for color discrimination. But if he grows up in a family of artists, all of whom can distinguish among five different shades, he might be considered unfortunate and genetically bereft.

When a gene's merits are judged by personal standards, generalization becomes difficult. Genes do not act in a vacuum; they cannot be judged out of context. A gene which is good in one person's view may be bad in another's, not because the gene itself changes, but because of differences in personal and social views.

Good News and Bad News

A gene can be both good and bad when it has several different effects, depending on the environment. No matter what your HLA type is, it contains potentially good and bad characteristics: it may be associated with a resistance to one illness, and a susceptibility to another. Among the cancers, for instance, the HLA markers Aw19 and B5 indicate that a patient has a better chance to survive at least two years if he contracts bronchial cancer, but is less likely to respond to therapy if he contracts Hodgkin's disease. This damned-if-you-do-and-damned-if-you-don't dilemma exists with most of the alternatives among HLA types.

Other markers present the same mixed blessing. Persons with high AHH levels have an increased risk of cancer after skin contact with some chemicals and a decreased risk for leukemia if the same chemical is taken orally. And among the blood group markers, people with type A are more susceptible than type O people to rheumatic diseases, but more resistant to stomach ulcers. Which is the good gene and which is the bad?

This does not imply that it does not matter which genes you carry. The whole point of knowing which genes you have is to help you decide what to avoid and what to enjoy. If a woman with blood type A is concerned about her susceptibility to forming blood clots while using oral contraceptives, she can elect to stop taking the pills.

Two-faced genes are more or less the rule in heredity. Most genes are compromises between desirable and undesirable characteristics.

When Bad Is Good

At the beginning of the nineteenth century, it was easy to decide between good and bad genes for *Biston betularia*, a species of moth that lives around Manchester, England. To protect themselves, the moths had developed a light grey coloring on their wings and bodies—a shade that blended perfectly with the rocks and tree trunks of the area. The thrushes, yellowhammers, and lepidopterists—all predators for the moths—used their eyesight to locate food and specimens. The better the moths were at looking like a part of their inedible, uninteresting surroundings, the better their chances of staying alive. The predominant genes for body color—the light shades of grey—had to be good genes; they helped the moths to survive.

The first black specimen of *Biston betularia* was snared in Manchester in 1848. It was probably a mutant strain, the result of two recessive, dark-colored genes appearing in the same creature. The insect didn't have a chance.

But over the next 50 years, black strains of the moth began appearing in larger and larger numbers. It appeared that the industrialization of England—with its soot and pollution—was changing the color of the moths' surroundings. The light moths stood out against the darkening background like lanterns in the night. Their predators picked them off with ease. The genes for grey body and wing color, so essential to the moths of 50 years before, were now dangerous.

The number of light-colored moths dwindled; the number

of dark moths increased. By 1895, about 98 percent of all moths living around Manchester were black. The same phenomenon was repeated in industrial areas of Germany, Poland, the Austrian Empire, and North America. The gene for black body and wing color had come into its own.

Now the pendulum seems to be swinging back. As we undertake the task of cleaning up our industrial messes, the trees and rocks are again becoming grey. Black moths are suddenly at a disadvantage. Once again, they are giving way to grey moths.

This process of evolutionary survival and genetic selection touches every living creature, including man. Many important genes have the capacity to be beneficial or harmful, depending upon the circumstances in which they are expressed. A particular genetic variation that proves undesirable in one time or place may be innocuous in another.

In the same way, traits that used to be critical to our own survival may no longer be so important. As man evolved from a hunter in primitive societies, he built a culture that could easily accommodate certain previously unhealthy genetic characteristics. Ten thousand years ago, a nearsighted hunter could easily become a target for other predators. Not only was he probably inept as a killer, but he was less apt to notice danger. Genes for poor eyesight, of which several hundred are known today, were considered bad; it is doubtful that those who carried them managed to survive very long under conditions which demanded daily visual acuity. But civilization and eyeglasses have changed that equation; nearsightedness today is an inconvenience to some and a modest cosmetic concern to others, but it is rarely a major disability, and it certainly is not life-threatening.

Resistance to malaria is another such example. Malaria is caused by a mosquito-borne parasite. It is still the most important infectious disease problem in the world. The parasite is found in South and Central America, Africa, and Asia, where *Anopheles* mosquitoes provide the transportation and susceptible human beings provide the food.

The parasites' banquet hall is the red blood cell, where they reproduce. In many cells, the production of new parasites continues until the cells burst. The victim suffers from chills, fever of up to 107°F, headaches, muscle pains, and anemia. Young children and pregnant women are particularly likely to develop symptoms that lead to death.

Yet despite the vast areas of the world in which the parasite is found, mankind has survived. Part of the reason is the high proportion of individuals who are genetically resistant to the disease. They can be placed into two groups: those resistant to the parasite *Plasmodium vivax;* and those resistant to the closely related *Plasmodium falciparum.*

In 1979, *P. vivax* was the agent for most of the 2,053 cases of malaria imported into the United Kingdom. In addition to immigrants and foreign tourists, about 13 percent of the cases occurred among sea or air crews and among English tourists and businessmen returning from abroad. According to the Malaria Reference Laboratory at the London School of Hygiene and Tropical Medicine, 89 percent of the cases contracted in Africa, India, and other parts of the world were of the *P. vivax* type. Nineteen seventy-nine was the seventh year in a row with a steady increase in the disease; and mutant strains of the parasite with greater resistance to the drugs used to combat it are beginning to surface.

There are, however, those who are protected from *P. vivax.* Those most likely to be resistant all seem to have a particular blood type known as "Duffy negative." They lack both of the two known forms of the Duffy blood type—small sites on cell surfaces that can be identified by a blood test. Although European and American Caucasians rarely carry the Duffy negative marker, practically all West Africans and approximately 70 percent of American blacks do.

The significance of the Duffy marker became apparent when Louis H. Miller and his colleagues at the National Institute of Allergy and Infectious Diseases in Bethesda, Maryland tested the ability of a parasite closely related to *P. vivax* to infect red blood cells. The infectious process could be followed under a microscope. When Duffy *positive* cells were placed on a slide with the parasites, the invasion occurred. The parasites clung to the cells' surfaces and burrowed through their membranes. But when Duffy *negative* cells were brought in contact with the parasites, the results were dramatically different. The parasites attacked, but they simply bounced off the cell membranes because there was no place for them to attach. The Duffy markers seemed to offer the parasites microscopic hooks on which they could hang.

In the late 1970s, Miller decided to test the theory about malaria resistance in Duffy-negative individuals with people who had actually been exposed to the disease. Experimental models were nice, but the acid test would come with a tryout in nature. The laboratory was the jungle of Southeast Asia. The subjects were black servicemen, mainly because blacks are split almost evenly between Duffy positive and Duffy negative, while whites rarely have Duffy negative blood. Miller knew that soldiers who had served in Vietnam had probably been exposed to malaria. He contacted the Medical Follow-Up Agency of the National Research Council, which keeps medical records of all veterans, and requested that the agency compile a list of every black soldier who had contracted *P. vivax* malaria and ask each one for a blood sample. To ensure privacy, the agency did not give Miller the names and addresses of the soldiers; to ensure that participation would be voluntary, the veterans were assured that their refusal to participate would have absolutely no effect on their veterans' benefits.

The agency's computer identified eleven cases of *P. vivax* malaria among the black soldiers. Miller tested their blood samples, and one after another came up Duffy-positive. The unanimity was extraordinary. Clearly, Duffy positives were susceptible and Duffy negatives were resistant.

But what about the more severe, life-threatening malaria caused by *P. falciparum*? In the 1979 British records, there were five malaria deaths, all of them in susceptible Britons who had visited African countries, all of them due to the *P. falciparum* parasite. For this type of malaria, the Duffy marker is useless; those who are negative are just as likely to contract the disease as those who are positive. But there are at least three genetic markers that can identify people who are more resistant to this parasite.

All three markers point to genes that we are used to thinking of as harmful: those that cause thalassemia (a fatal blood disorder), sickle-cell anemia, and G-6-PD deficiency. The genes in question, however, trigger the most virulent forms of the diseases only when they are inherited in pairs; and, in each case, it seems to take only one gene to provide some protection against malaria. Therefore, many who have avoided the ravages of malarial infection can thank their "defective" genes.

The first person to suggest that a connection exists between these genes and resistance to malaria was the great British geneticist, J. B. S. Haldane. Haldane pointed out that if these genes were always harmful, they would long ago have been eradicated in the process of natural selection, as those who carried them died without reproducing. Because so many people in the Mediterranean area retained the thalassemia gene, and because that same region had a high incidence of malaria, he theorized that people who carried the thalassemia gene somehow *benefited* from it by being resistant to the malarial parasites. Of course, the protection helped only those with a single copy of the gene; those who inherited two copies invariably came down with the blood disorder and died.

Haldane's hypothesis made good common and scientific sense. To test it, other scientists began to survey Mediterranean populations for the relationship between gene and resistance. They discovered not only that it existed, but that it applied to sickle-cell genes as well. In areas where malaria is common, children who are carriers of the sickle-cell gene have fewer parasites in their blood than children with normal genes, and fatal cases of malaria rarely involve carriers of the sickle-cell trait. They also discovered that if they drew two maps—one which depicts the areas where malaria tends to be fatal and one which outlines locales where most sickle-cell carriers are found—they can almost be superimposed upon each other. Today, little doubt remains that carriers of the sickle-cell gene are more resistant to malaria than those without it.

The third major resistance gene, for G-6-PD deficiency, is also common in malaria-infested areas. And scientists in Nigeria recently showed that red blood cells from carriers of G-6-PD were more resistant to parasitic invasion. Normal blood cells were invaded up to 80 times more efficiently than the cells with the G-6-PD gene.

When an individual carries two markers for resistance, the combined resistance seems to be greater than if only a single marker is present. In Sardinia, the incidence of malaria is significantly lower in people who carry both the G-6-PD deficiency and the thalassemia markers than in those with only one of them.

The question of whether a gene is good or bad need not always be rhetorical. It becomes critical if we take into consideration the conditions under which it must function. In the high altitudes of Boulder, Colorado or the Himalayas, single copies of a gene for malaria-resistance such as the sickle cell trait may turn out to be dangerous because of their carriers' higher demands for oxygen. In Nairobi, Kenya or Caracas, Venezuela, that same gene can save lives. Almost without exception, deciding whether a gene is good or bad depends upon when, where, and how it is expressed.

The Spice of Life

For human beings, as well as for every other species, the key to sustaining our existence as a population is to survive and flourish within many different environments. And the way we survive is by adapting. Genetically, that means that our genes must change over the generations as the environment changes, so that our species can maintain a selective advantage. It also means that the genetic variants that we all carry within the surplus genetic material in our chromosomes probably act as a reserve pool from which we can draw when our regular sets of genes cannot respond. Diversity among inherited traits is a natural way of ensuring that our species will continue to flourish. For genes, variety is the spice, and the essence, of life.

Clearly, our species retains several alternatives for many traits. The ABO blood groups, for instance, come in four different varieties: A, B, AB, and O, each characterized by different blood proteins. In Southern England, 44 percent of the population has blood type O; 44 percent has blood type A, 8 percent has type B; and 4 percent has type AB. Apparently, types O and A must have conferred a selective advantage over the other types. As a result, they were passed on more often to offspring who survived, and their incidence in the population increased.

This same method of selection probably occurs in ABO blood groups throughout the world. Type O blood predominates among the Indian populations of Central and South America, a fact that, some years ago, sent scientists in search of a disease that might have selectively killed off those with other types. Almost from the very beginning, they suspected syphilis. Many medical historians believe that syphilis did not exist among the relatively isolated populations of the New World before the arrival of Columbus. When the Spaniards finally did import it from Europe, the more susceptible individuals and their progeny probably began to die off. We know now that women with syphilis deliver stillborn babies; we also know that the immune systems of people with blood type O respond more effectively to syphilitic infection than the immune systems of other blood types. Taken together, these facts suggest that people with the O marker were often spared the disease, while others were probably not as fortunate.

Similar theories explain why the other blood groups still exist. The mosquitoes which carry the viruses for yellow fever, encephalitis, and other human diseases prefer people with blood type O. Those with type A and B blood therefore tend to have an advantage when mosquitoes are spreading disease through a population. Similarly, infants with blood group A were more likely to survive some of the epidemics of infant diarrhea that used to sweep Central Europe. Each blood group manages to confer some advantage to those who carry it. And the incidence of each group in different parts of the world today probably reflects the usefulness of its advantages.

The fact that a bad gene can become good simply because of a change in the environment provides a clue to many medical puzzles. One of these is the number of people who inherit a tendency for diabetes. In 1980, there were about 10 million diabetics in the United States—one person in twenty. If a gene's survival depends upon its ability to confer a selective advantage, how could such a trait persist?

Diabetes was first described in an Egyptian papyrus around 1500 B.C. It is characterized by a person's inability to use sugar properly. Whenever a diabetic eats sugar-laden food, the quantity of sugar carried by the blood increases abnormally because of a defect in the regulatory mechanisms. The excess sugar can lead to kidney disease, gangrene, and heart disease. Diabetes is also the second leading cause of blindness in the United States.

Diabetes occurs in two forms: adult-onset diabetes, which generally appears after the age of 40 and in obese people; and juvenile diabetes, which usually appears in people under the age of 20. Adult-onset diabetes is the more common form. It tends to surface gradually and run in families. In fact, if two parents are diabetic, their children are almost certain to develop diabetes if they live to 80 years of age. Twin studies confirm the genetic basis of the disease. In one study of over 100 pairs of identical twins, whenever one twin developed diabetes after the age of 50, the other almost invariably did too, within a few years. Unfortunately, no good marker yet exists that can predict adult-onset diabetes; family records are still the only useful tool.

Until Pablo Rubinstein discovered the connection between HLA type and *juvenile* diabetes, the role of genes in that form of the disease was uncertain. The onset of juvenile diabetes tends to be abrupt, as if it were caused by some environmental factor. It is also known that certain chemicals can destroy the pancreatic cells that produce insulin.

But Rubinstein discovered that HLA markers in families could predict whether the brothers and sisters of a diabetic child had a 1 in 1000, 1 in 50, or 1 in 2 chance of contracting the disease, a far more accurate measure of risk than the 1 in 8 chance that every sibling has if HLA type is not taken into account.

Why the highest risk is only 1 in 2 is still a mystery. But it is a mystery with some exciting clues. For years, physicians have noticed that viral infections in children can be followed by diabetes. The connection has occurred too frequently to be explained by chance. Finally, in 1979, Ji-Won Yoon, Takashi Onodera, and Abner Notkins, at the National Naval Medical Center, found the first solid evidence to support the link.

A healthy 10-year-old boy suddenly developed diabetic symptoms three days after contracting a flulike illness; 7 days later he died. The researchers isolated a virus, Coxsackie virus B4, from his pancreas. If this virus had a special capacity to cause diabetes in humans, they reasoned, it might also cause diabetes in mice. Certain strains of mice are known to be genetically susceptible to diabetes, while others are resistant. When the investigators inoculated the two strains with the Coxsackie virus, the susceptible mice developed diabetes; the resistant mice did not. Genetic susceptibility and environmental trigger were present both for the mice and for the boy. (Coxsackie virus may turn out to be a common trigger for diabetes, but several other candidates are also being tested.)

We now know that genetic susceptibility plays an important role in the development of both juvenile and adult-onset diabetes. Yet despite their severe consequences, the gene or genes involved have been with us for thousands of years. Why?

One possible explanation grew out of experiments carried out by Douglas Coleman of the Jackson Laboratory in Bar Harbor, Maine. In 1979, Coleman took both normal mice and mice that carried a gene for susceptibility to diabetes and gave them access to identical sources of food and water; they could eat and drink as much as they liked. After one week he cut off their food, while allowing them to continue drinking. The results were astonishing. The mice which carried the diabetic gene lived 23 to 46 percent *longer* than the normal mice. The gene apparently permitted its carrier to use the food it had stored more efficiently than normal mice could.

Coleman calls the diabetic gene a "thrifty trait." It appears to function in the wild as well. When animals from desert or semidesert areas, such as the Egyptian sand rat or the South American rodent tuco-tuco, are transferred to captive environments with plenty of food, they tend to develop diabetes.

If the same mechanism operates in humans, the diabetic gene may have an advantage in areas of the world with limited food supplies. In times of famine, individuals with the gene may have been more likely to survive. But when a population that is used to sporadic supplies of food is provided with a more or less continuous source, the diabetic tendency can become epidemic. This has been one explanation for the high rate of diabetes among the Seneca Indians in New York State. The disease is probably the most common chronic health problem among those on Seneca reservations today.

When a disease as debilitating as diabetes can have its beneficial side, we begin to see the degree to which genes depend on the environment. The genes that predispose us to diabetes and other illness persist precisely because we live in a constantly changing world. They are an integral part of the struggle for survival. They are one way of making sure that the species will prevail, regardless of the environment.

Genetic Lemons

Every group has its own particular distribution of genes that make it susceptible to some diseases. Among American Caucasians, one out of 25 people carries the gene for cystic fibrosis. Greeks have an increased incidence of Cooley's anemia, Jews of Tay-Sachs disease, Armenians of familial Mediterranean fever, South Africans of porphyria, French Canadians of tyrosinemia, Chinese and Thais of alpha thalassemia, and Finns of about 20 otherwise rare genetic diseases. And every individual in every group carries genes that would prove undesirable under certain circumstances. We are, in short, all carriers. We all have the potential to pass on deleterious genes, even though the odds of our marrying someone from among the general population who carries the same genes are remote. As we reproduce, we play a genetic dice game, with the odds enormously in our favor, but with potentially harmful consequences for our children if we should lose. Fred Bergman of the National Institute of General Medical Science has summed it up more succinctly: "In one way or another, we are all genetic lemons."

Genetic prophecy offers us a chance to increase the odds against our children unknowingly suffering from genetic defects that might predispose them to disease. But the game continues. What we must try to discourage is the idea that our avoiding genetic pitfalls somehow makes us superior to those who do not. No matter how much the dice may be loaded in our favor, an unlucky roll has nothing to do with the worth of the player.

PART III
BEHAVIOR

8

THE FALSE CONTROVERSY

A devil, a born devil, on whose nature nurture
can never stick.

PROSPERO describing Caliban
The Tempest, William Shakespeare

A mother looks at her child and sees the mirror image of her own grandmother.

A lean, lanky baby is born to parents of average build. Without the assistance of an overactive thyroid, he grows to a size that makes professional basketball scouts slaver.

Identical twins are kept together in school. They look alike, yet somehow it is easy to tell them apart. One is extroverted, active in sports; the other spends her time in the library, lost among the books.

Inherited or acquired. Nature or nurture. Gene or environment. The two have been the center of a conundrum that has embroiled religious and scientific thinkers since the publication of Charles Darwin's *The Origin of Species* and has intrigued armchair philosophers for almost as long as there have been coffeehouse debates.

From the moment the debate began, support has swung from extreme to extreme, according to the dictates and fashions of the times. Around the turn of the century, scientists realized how much easier it was to judge the inheritance of characteristics on the basis of the color of a fruit fly's eye than on the passing of a complex behavioral trait from parent to child. Simple physical characteristics were so clearly genetically influenced—and the idea that behavior was inherited was so socially acceptable—that everything that could be called a trait was labeled genetic. One group of scientists claimed, for instance, that "alcoholism, seafaringness, degeneracy, and feeblemindedness" were each due to single genes, and that the children of unions between black and white would have "the long legs of the Negro and the short arms of the white, which would put them at a disadvantage in picking things off the ground." The terms *eugenics* was coined, and societies were set up to encourage young people of "good" stock to breed freely, so that the quality of the human gene pool would improve. In the United States, restrictive laws of immigration were passed to slow the flood of the "inferior" peoples of Eastern Europe.

But even as these bizarre ideas reached their logical conclusion in the racial beliefs of the Nazis in Germany, the pendulum had begun to retrace its arc. Sigmund Freud and Carl Jung were in; fruit flies and Gregor Mendel were out. The lure of genetics as an explanation for the development of all kinds of traits began to fade; the ways in which people were fed and sheltered, reared and educated—in other words, the environments they encountered—were accepted as the main determinants of complex emotional characteristics. "There is no evidence of the inheritance of traits," claimed John B. Watson, the father of behavioral psychology. And his colleagues set out to try to prove him right.

The widespread rejection of the role of genetics was not fueled entirely by hard evidence, or even by the irresistible attraction of better ideas. Good scientists are human beings as well; their natural tendency was to disown the theories that contributed to the atrocities of the Nazis and the mishandling of minorities in the United States. Now we know that they discarded some good ideas with the bad and pushed the trend toward an environmental explanation of traits too far. It was inevitable that the pendulum would swing back again, toward a rational, scientifically acceptable middle ground. And swing it did. Today, supported by a better understanding of the relationship between nature and nurture, it hangs squarely in the middle, as scientists continue to develop evidence that attests to the power of each.

We now know that genes and the environment do not compete with each other. In some situations, of course, either one or the other can be extraordinarily powerful: Genes have no way of preventing someone who is standing at the site of a nuclear explosion from being vaporized, nor can the environment give life to a baby born without essential chromosomes. And the precise and unambiguous genetic blueprints that mandate that humans should walk upright, or, for that matter, that some should have blue eyes instead of brown, can only be countermanded by extreme environmental conditions, such as accidents, perhaps, or congenital birth defects.

Generally, however, one factor cannot completely dominate the other. Instead, gene and environment are complementary; they work together to produce a final result. The phenotype of any one person—the composite of all the characteristics that he or she carries—is a product of the *interaction* between gene and environment.

Why do heredity and the environment not compete? Primarily because they have assumed entirely different tasks. The genes are like the tools of a painter's trade: the lighting in the garret, palette, brushes, and paints. They cannot operate in a vacuum. They require external catalysts before they can function.

The environment provides the needed boost. It constitutes the artistic effort itself. It takes the proffered genetic tools and works with them to produce the finished portrait. Like the artist, it operates within a range defined by those tools. For no matter how powerful it is, the environment cannot force genes to manufacture products that they were not designed to manufacture.

The difference between seeing heredity and environment as competitors and recognizing their supportive relationship is a critical one, not only in understanding the concepts behind genetic prophecy but in realizing how we, as living organisms, adapt in order to survive in a changing world. For it is the design of nature never to work alone; and it is the nature of nurture always to act within previously described boundaries.

The theory of this relationship is quite straightforward. In practice, however, the situation is often more complex. Genes, for instance, often act in concert with other genes. They can be influenced, altered, modified, triggered, or shut down by the actions and products of their neighbors. As such, they not only help control the impact of the environment, but actually

constitute a part of that environment for the genes they
influence.

Genes can affect other genes in various ways. Suppose, for
instance, that each of us inherits one of two genes: the first
makes us particularly susceptible to arsenic poisoning; the
second confers resistance. The dose needed to poison some
one who has the gene for susceptibility might be so small that
it has no effect on someone who is resistant. But what if the
gene for susceptibility is not the *only* one involved in determining arsenic poisoning? What if that same susceptible individual also inherits a gene whose product interferes with the
ability of the intestines to absorb arsenic into the bloodstream? Clearly, the vulnerable tissues may never even be
exposed to a dangerous dose. And if someone who is resistant
happens, at the same time, to inherit a gene which *enhances*
arsenic absorption, we might find ourselves with a medical
paradox: The person who seems resistant becomes ill; the one
who seems susceptible is perfectly safe.

The importance of uncovering the *total* effect of the genes
for any one condition, and not just the impact of the genes
that are directly involved, is critical to genetic prophecy. The
more complex traits we carry—from height to behavioral
characteristics—are seldom influenced by a single gene. Most
are the products of constellations of genes working together
to form the raw material that the environment molds. Even
more important, the influence of these genes is different for
nearly every trait. In some, the genes play a major role in
dictating a narrow range of responses among which the environment can choose. In others, the genes allow for a wide
variety of responses. In the first case, the genes appear to
"dominate" because the environment seems to have little or
no effect on their expression. In the second, the environment
appears to "dominate" because each different environment
seems to elicit a distinctly different genetic response.

The relationship between the two factors does not spring to
life only when particular genes meet specific conditions. It
takes place over time, in a constant give-and-take as we search
for the perfect balance between the two. A particular set of
genes may enable us to modify the environment to make it
more conducive to our survival. We may, for instance, popu-
late the temperate zones of our planet, learn to combat
excessive periods of heat with air conditioning, and avoid as
best we can certain types of pollutants. At the same time, the

environment may be stimulating genetic change by influencing the process of natural selection, by helping to "weed out" genes that are less useful, less productive, more debilitating. The "environmental" causes of cancer, for instance, exist primarily because the environment is changing so quickly. Our genetic heritage, fashioned to confront one type of environment, has not yet learned how to handle the changes; the result is illness. In the future, the environmental changes might begin to help eliminate deleterious genes. The incidence of cancer may some day decline simply because, genetically, we are learning to respond to and neutralize environmental insults.

This seesaw relationship continues. Genes, unlike leopards, can and do change their spots; and environments vary widely in their influence. The interaction between them depends upon the relationships among all sorts of traits, pressures, and responses. It is, in short, a microcosm of human society itself, a composite of individual elements and groups, all of whom, together, paint the final picture.

The Salamander's Solution

Because of this complicated arrangement, we often have difficulty pinpointing where genes leave off and environment begins. To make matters more complicated, except for the simplest characteristics, such as eye color or single-gene diseases like hemophilia, we do not even inherit actual "traits" from our parents. Rather, we receive specific sets of molecules which express themselves according to the environments into which they are introduced.

Early scientists had no reason to know this. They were often baffled by the tricks that nature seemed to play. Zoologists were especially fascinated by the case of the Mexican salamander, which, under different conditions, assumes utterly different forms. An embryo placed in water develops gills, a ponderous body, and a tail designed for swimming. Its twin, raised on land, breathes air and evolves a smaller, lighter body. The two carry identical sets of genes; but under different environmental pressures, they develop in ways so dissimilar that, at first glance, they seem to come from unrelated species.

While all forms of life undergo these kinds of adaptations, the environmental factors are seldom as dissimilar as they are for the Mexican salamander, and the changes they instill are not nearly as obvious. A person's height, for instance, is closely linked both to his genetic heritage and to the environment in which he grows. Genetically, tall people tend to be born into tall families and short into short. The same holds true for entire populations as, for example, in the seven-foot giants of the Ibo tribe of Nigeria or the short, stocky people of Southeast Asia.

Research has shown that average stature in a population tends to remain more or less constant. When it increases, the reasons turn out to be mainly environmental; improvements in nutrition and health during the first years of life seem to have the most significant effect on one's height in later life. In Switzerland, for example, where records of the stature of conscripts into the army have been kept for centuries, researchers have found that, during the nineteenth century, intellectuals, merchants, and students were generally taller (by 2 to 5 inches) than factory workers, farmers, and blacksmiths. The differences in height followed the pattern of class distinctions that existed in Switzerland at that time. They were largely influenced by the relative quality of the food and health care that the conscripts had received while growing up.

By 1930, class distinctions had begun to disappear; the environmental factors affecting height had begun to even out. As a result, factory laborers, who had measured a solid 5 inches shorter than intellectuals in 1887, gained an average of nearly 6 inches in stature. The lot of the intellectuals had improved as well, but the differences between their lives in 1887 and 1930 were not nearly as great. They grew an average of an inch and a half taller. So while they were still a shade taller than the factory laborers, the extreme differences found in the past had disappeared.

The size and complexity of the Swiss population indicates that the increase in stature was not simply a matter of everybody within a single group reaching his potential, but rather a question of various groups with different potential heights all reacting to environmental conditions. Underfed laborers in the tallest group might well end up the same height as healthy intellectuals in the shortest group; and middle-sized merchants might stand eye-to-eye with larger blacksmiths. The change in the *average* stature was a product of improvements in the environment, but *individuals* grew only within the ranges permitted by their genetic heritages.

The importance of the genetic contribution to size cannot be overlooked. The most obvious genetic trait that we all have is our sex. The difference in height between men and women in our well-fed, generally middle-class society averages about 5 inches. This is a powerful and significant difference that points directly to the genes.

Inside the Cell

Height is a multifactorial trait—that is, it is influenced by a broad range of genetic and environmental potentials. That same kind of balance exists throughout biology. Scientists have traced it right down to the molecular level, to the regulation of a single gene's production inside a microscopic cell.

The bacterium *E. coli* is a tiny, usually harmless inhabitant of the human bowel. Because it manages to survive in all kinds of unpleasant surroundings, multiplies with little difficulty, and eats practically anything, it has been an experimental favorite of researchers for years. In fact, scientists have worked so extensively with it since its discovery in 1885 that they now can identify and characterize well over half of all *E. coli*'s chemical properties, making the promiscuous little bug by far the best understood outpost of life on this planet.

In the early 1960s, Francois Jacob and Jacques Monod, two French molecular biologists, undertook a study of how *E. coli* regulated the production of its genes—how, in other words, *E. coli* managed to start and stop the manufacture of its various enzymes and structural proteins on cue. They decided to use as their model a single identifiable gene among *E. coli*'s 2,000, one that produces an enzyme, called beta-galactosidase, which helps the cell break down lactose, a sugar that provides it with energy.

The gene that directs the production of beta-galactosidase functions only part of the time. It does not funnel a constant stream of enzyme into the bacterium, but is somehow turned on only when lactose is present. Clearly, *E. coli* has evolved a kind of "molecular wisdom," a system that recognizes the presence of lactose and responds accordingly.

It is easy to invent fanciful scenarios to explain how *E. coli* might turn its enzyme production on and off. But Jacob and Monod found that the cell's actual method is simple and efficient, and, for molecular biologists, almost too beautiful for words.

Jacob and Monod discovered that the system that produces beta-galactosidase actually consists of two genes: one that contains the code for the production of the enzyme itself, and another, called the "repressor" gene, that oversees and controls production. The repressor gene, which sits close by the beta-galactosidase gene, produces a constant stream of a "repressor" protein. But the protein can only prevent the enzyme's production part of the time; at other times, it does not.

After careful searching, Jacob and Monod managed to uncover the bacterium's secret. They discovered that the repressor protein has a special affinity for a small section near the beginning of beta-galactosidase's gene. After it is produced, the protein drifts around until it finds its natural resting spot. And there it sits, acting as a brake on the machine, shutting down the beta-galactosidase factory completely.

If the repressor protein simply remained there all the time, however, *E. coli* would never produce a supply of beta-galactosidase. That arrangement may be perfectly fine for some genes in human cells, where genes involved in creating, say, a muscle cell are shut down permanently in cells which are destined to become parts of nerves. But *E. coli* needs its enzyme to survive. As a result, it has found a way to lift the siege of the repressor protein when lactose appears.

By something more than coincidence, the lactose that enters *E. coli* finds repressor proteins very attractive. The sugar molecules latch on to the protein and pull it from its seat on the bacterium's DNA. When the protein is gone, the beta-galactosidase gene begins to produce its enzyme; and it continues to produce until the enzyme has broken down all the lactose and the repressor proteins are again free to settle and prevent the gene from functioning.

The beauty of this circular system is obvious. *E. coli* cannot afford to waste energy on the useless manufacture of an enzyme, but it needs the energy that the enzyme helps extract from sugar. Therefore, it blocks production with a molecule that the sugar itself is attracted to. The presence of

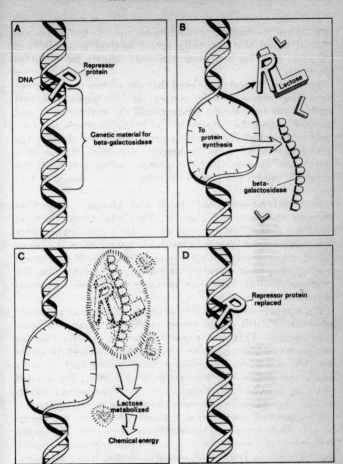

3. Gene and environment in the cell. (A) The repressor protein prevents the beta-galactosidase gene from beginning the production of the enzyme. (B) Lactose enters the cell; the repressor protein is lifted off the tangle of DNA, permitting the synthesis of beta-galactosidase to begin. (C) Beta-galactosidase breaks down lactose, providing the cell with energy. (D) With no lactose to lift it free, other repressor proteins attach to the DNA, preventing the production of beta-galactosidase.

the sugar "switches on" the beta-galactosidase gene (by removing the repressor protein) and triggers the production of the enzyme. When all the sugar has been converted, the blockage is reinstated. The two genes—the repressor gene and the enzyme's gene—are there all the time. *But, by its presence or absence, the sugar determines when they act.*

The production and suppression of beta-galactosidase is a perfect example of the way genes often depend upon environmental factors to function. For lactose is an external element that enters the cell. It triggers a gene's production and the consequent yield of energy simply by its presence. Without it, the entire sequence would never take place, and the very nature of *E. coli* would change.

Salamanders, stature, and *E. coli* illustrate the relationship between genes and the environment for purely physical characteristics. But as scientists are now discovering, the same pattern exists for behavioral traits. Again, the genes determine the range of possibilities and the environment chooses from among them. The main problem with discussing behavior in the same way we talk about physical traits is that our understanding of what constitutes an element of any complex behavioral pattern is still a matter of conjecture. What we know of the mechanisms of behavior is primitive compared to our knowledge of, say, genes and disease.

Still, hints of the behavioral influence of genes have been around for decades. People who are born with phenylketonuria (PKU), for instance, grow up mentally retarded. Until 1934, they were indistinguishable from others who suffered retardation. Then, Asbjorn Fölling, a Norwegian physician, tested two mentally retarded sisters and found that their urine turned green when it was added to ferric chloride—a reaction that indicated a biochemical abnormality. Later, other researchers discovered that others with the same problem lacked a single crucial enzyme that prevented the conversion of one amino acid, phenylalanine, into another. The build-up of phenylalanine ultimately causes brain damage.

Five years after Fölling's discovery, G. A. Jervis surveyed over 20,000 people in the United States who had been institutionalized because of mental retardation. He found that nearly 1 percent had PKU.

Fölling's test provided a way for physicians to know in advance that people in a certain group would become mentally retarded, for it tested, indirectly, the absence of an enzyme—a genetic marker. But what good would it do? PKU was still uncontrollable. Identifying those at risk meant nothing unless a way could be found to prevent them from succumbing to their genetic heritage.

In the mid-1950s, researchers found what they were looking for. A diet low in phenylalanine was discovered to be capable of preventing the biochemical changes caused by PKU. In 1958, the first commercial preparation of the low-phenylalanine diet was made available. Since then, newborn infants who would have been headed irreversibly toward an institutionalized life have been saved from mental retardation by a simple test and a carefully planned diet.

Finding the underlying cause of PKU was an important milestone on the behavioral side of genetic prophecy. It established a firm biochemical basis for at least one kind of mental abnormality. It offered a genetic marker that could be used to predict who would become ill. It implied that understanding the biochemical and genetic nature of other so-called mental illnesses might also prove effective in prediction and protection. And it helped catalyze the movement away from the strictly environmental interpretations of behavior that were then in vogue. Today, screening newborn infants for PKU is a required procedure in most states.

The history of our growing understanding of PKU is a perfect example of how much we depend on an ever-changing state of knowledge to support the assumptions on which we act. PKU is caused by a defect in a single gene. It is triggered by the presence of a clear-cut environmental agent. How thorough is our dependence on information that might change at any time? Try for a moment to answer the question: Is PKU a genetic or an environmental problem?

Before 1934, PKU did not even exist as a separate entity. It was lumped under a broad heading that included dozens of different abnormalities, all of which resulted in mental retardation. Then its genetic connection was uncovered. It was considered a disease that was beyond our control, a predetermined condition that medicine could do nothing about. Finally, researchers isolated its environmental component, and our perceptions had reached a point of synthesis. Certainly, those who lacked the enzyme were *predisposed* to retardation. But their fate was far from certain. We learned to provide them with protection against what had been considered first a "mental," and then a "genetic" problem. Full understanding led to control. By a simple change in diet, people with PKU could be lifted from the ranks of the disabled.

Mirror Images: The Twin Saga

Asbjorn Fölling was lucky. He found two people with the same condition, managed to conduct the right test, and read his results correctly. But he was working with a simple problem: a single genetic defect that triggered a clear and obvious abnormality. What happens as we deal with other behavioral traits? How do we go about evaluating humans for those traits in ways that are reliable, scientifically acceptable, and safe for the people being tested?

Behavior itself is perhaps the greatest mystery, the least explored realm, left to us in biology. It resides in the least understood organ in the body, the brain. It consists of a confounding number of vague, poorly defined characteristics. It is a complexity surrounding complexities, an unknown composed of unknowns. It is, in fact, so difficult to define that behavioral scientists often disagree on the simplest division that can be made—between normal and abnormal behavior.

This confusion is a natural element in the exploration of human behavior, a search that is only just beginning. But humans are only the tip of the behavioral iceberg. The same debates go on among those who study more elemental forms of animal life. They observe discrete examples of behavior and ask themselves: Conscious or unconscious? Learned or inherited?

Examples of this discord are everywhere. There are those who cannot decide, for instance, whether the food dance of the honey bee constitutes a form of thought or merely an inbred reaction to environmental stimuli. Every new piece of evidence complicates the question. Honey bees, it seems, have far greater control over what they tell other honey bees than we ever imagined. Whenever they locate food, they use rays of polarized light to pinpoint the spot. Then they orient themselves by that same source of light when they return to the hive to dance, telling their fellow bees the distance, direction, and even the desirability of the flowers they have found. The sun, of course, is the source of their polarized light. But the sun moves; by the time a bee scout has negotiated her way back home and begun her dance, its position in the heavens has changed by an amount that is small but significant. Yet somehow the bees adjust. They can recognize the different angle of light, calculate the degree of dispersion, and incorporate it into their story. Their ability to do this is not simply a matter of instinct: Mature, seasoned bees adjust better and make fewer mistakes than younger ones.

The implication that communication among bees depends partly on a learning process was in itself a startling revelation to those who had supposed that everything in a bee's behavioral repertoire was predetermined. And so they began to experiment more fully. Scientists asked: Do bees actually fix on the sun, or is their response more primitive than that, a mere matter of timing? They took hives of bees from the northern hemisphere (in which the sun appears to move from left to right) and transported them to the southern hemisphere (where the sun appears to move from right to left). They found that the sun's reversal causes the bees to make mistakes, but only for a while. Then they adjust to the reversal and change their dances to take into account their new environment. Scientists then asked: Can bees anticipate? They took a source of food and moved it a measured and even distance farther away from the hive every day. To their astonishment, the bees learned to adjust to the movement, flying past the area where the food had been the day before, congregating at the spot where it would logically be placed next.

What is going on here? How can bees recognize and adjust to artificial changes in their environment? When they have always seen the sun moving in one direction, how can they track it so quickly when it moves in another? When they have never confronted a flower that walks, how can they actually learn to guess that a food source is going to move, and be waiting for it when it arrives?

The riddles posed by simpler animals illustrate how difficult it is for us to define logically the elements of human behavior. Furthermore, with humans, the kinds of testing we can do with bees is not possible. If humans are poor specimens for laboratory tests that involve disease, they are even less suited to testing for behavioral traits. The tangled threads of heredity and environment are so tightly interwoven that an enormous amount of control over the testing process is needed before we can get some answers. And that amount of control is simply not possible.

The idea, then, is to find a way to cut down *naturally* on the variables, to limit them so that we can study a few at a time and reduce the enormous, unwieldy dilemma that surrounds the complex set of actions we term "behavior" into smaller, more calculable, constituent parts. We have been able to do just that with animals, by breeding rats, cats, guinea pigs, mice, dogs, and monkeys so that each individual within a breed or strain becomes essentially identical to all the rest. With identical animals, we can vary the environments to which they are exposed to discover how environmental variables choose among the options offered by the genes. We can subject them to extremes with drugs, heat, light, noise, chemicals to evaluate the limits of their tolerance. We can even modify their genes by careful breeding, creating animals only one or two or three characteristics removed from others.

We cannot do these things with humans. But we do have one weapon in our battle to discover the roots of behavior. We have at our disposal human clones—more than 10 million sets of them around the globe, each carrying precisely the same complement of genes as one, two, or three of the others.

Twins are born about once in every hundred births. Approximately two-thirds of them are fraternal twins, children born of the same mother at the same time but formed from two separate ova fertilized by two separate sperm cells. Fraternal twins are not clones. Their different sets of genes mark them as dizygotic twins, or twins born from separate fertilized eggs. Fraternal twins do not even necessarily have the same father, a fact dramatically illustrated by a German woman who gave birth to two sons, one black and one white. Were it not for the time they share in the womb, fraternal twins would be just brothers and sisters, as alike as siblings born under more conventional circumstances.

The other third, however, consists of identical twins formed from a single fertilized ovum that has divided once and split asexually into two separate cells. The cells grow, each into a separate fetus. But because they sprung from a single cell, these twins carry exactly the same set of genes. They are, in fact, clones of each other. Hunting dogs that can tell the difference between the body odors of fraternal twins cannot distinguish between the odors of identical twins. Any differences that do crop up between them are the result of the influence of the environment.

Sir Francis Galton, a cousin of Charles Darwin and the man who coined the word *eugenics*, was the first to recognize the potential value of twin studies. In 1875 he published a monograph titled "History of Twins as a Criterion of the Relative Powers of Nature and Nurture," suggesting that the close links between twins might make it easier to trace hereditary traits. He spent most of his life working with families and twins and trying to unravel some of the complexities his studies revealed.

Although Galton intuitively hit upon the idea that twins can provide valuable information about the ways genes and the environment work together, he had no way of distinguishing between fraternal and identical twins. As we now know, the fact that two people look and act alike is not enough to confirm their genetic identity; neither is the existence of one or two placentas at the babies' birth—a criterion some scientists used to use—since some identical twins have separate placentas, while some fraternal twins' placentas are fused and indistinguishable. Differentiating between identical and fraternal twins was, in fact, impossible before the 1920s, when physicians found that blood groups and proteins could provide solid, if not incontrovertible, evidence of genetic identity.

During the first twin studies, researchers were most amused and charmed by the *similarities* they found among their test pairs. Simply recognizing that these poeple had identical genetic structures gave them dozens of new options for studies: They could examine the twins in isolation. They could compare them to fraternal twins—with different genes, the same births and environments. Or they could chart them against normal siblings, in whom genes and birth times are different while the environment is similar, to evaluate the impact of identity on physiology, behavior, and personality. Identical twins gave them some control over a few variables. And so the twin studies poured forth.

But criticisms of these evaluations mounted. People found, for instance, that "identical" twins are not always exactly identical. One of a pair can have Down's syndrome, an extra chromosome which leads to mental retardation. And pairs of identical twins have even been of different sexes: one with the full XY—or male—complement of sex chromosomes; the other, having lost the "Y" chromosome by an error in chromosome division when the original single cell first divided, with a chromosome configuration called "XO," which causes the fetus to develop thoroughly female characteristics in a condition called "Turner's syndrome."

But the most important shortcoming stemmed from the assumption that identical twins grew up much as other children did and could be judged in the same way. Critics pointed out that subtle differences in natal and prenatal conditions could account for some differences between identical twins; they argued that parents often react to similarities either by reinforcing identical traits or by demanding that the children strive to be different; and they noted that identical twins often seem to divide the two sides of a complete personality between them, with one becoming the active, aggressive half while the other accepts a passive role.

The advent of highly reliable blood tests and the results of studies which showed that the twins' environments, both before and after birth, had little effect on their degree of similarity defused many of the attacks. But confounding factors were difficult to measure; the criticisms were impossible to disprove completely. And so, while the studies continued, the findings they generated were seldom accepted as absolute.

There was, however, another avenue of research. Some pairs of identical twins are separated at or near birth and raised apart, each uninfluenced by the other, in different environments. Finding them was like looking for two identical grains of sand on a beach; but, in 1937, the first study was published by two scientists, who, in 10 years, had managed to collect only 10 such pairs.

Over the next 40 years, researchers managed to study about 80 pairs of identical twins reared apart. The sample was tiny; the tests offered them were never the same. The conclusions were regarded by most psychologists and physicians as quaint and unsupportable. But some could see that these were twin studies with an important twist: They removed the environmental influences that might affect twins raised together; they offered an opportunity to pinpoint the *differences* among people with identical sets of genes, a chance to determine exactly what effect environmental variables could have.

In 1979, a break came, and from a completely unexpected

ource. Jim Lewis, a steelworker from Elida, Ohio, searched or and found Jim Springer, the twin he had been separated rom since 1939 when they were four weeks old. The two met t Jim Springer's house in Dayton with their families close eside them for support. When they finally faced each other fter nearly 40 years, they shook hands stiffly. Then they urst out laughing. They sat down and poured the champagne. They pointed out the differences between them: pringer wore glasses and combed his hair over his forehead; Lewis used nothing to correct his vision and brushed his hair o the side. But Springer took off his glasses; Lewis pushed is hair forward; and sitting side-by-side, they looked like hree-dimensional xeroxes off a good copy machine, two of a ind.

Two weeks later they met again, at Lewis' home. A local eporter heard of the meeting and had an article reprinted in he Minneapolis *Tribune*, where it was spotted by a psychologist at the University of Minnesota, Thomas Bouchard. Bouchard had been conducting twin studies for a decade; he new an opportunity when it strolled over to him and laid its ead in his lap. Within hours he had petitioned the University for funding. Within two weeks he had paid the twins' assage to Minneapolis for a full seven days of medical and sychological testing.

Bouchard is a psychologist, not a geneticist. He has been rained to think in terms of the environment, to assume that he critical factor in personality and behavior is the rearing of child, to place the genes in a subsidiary role. What he ound with the "Jim twins" was extraordinary:

- Both preferred math in school; both hated spelling.
- Both took law enforcement training. Both engaged in similar hobbies—mechanical drawing, building miniatures, carpentry.
- Since their youths, they had both vacationed along the same 300-yard strip of beach in Florida, driving there and back in Chevrolets.
- Both married and divorced women named Linda.
- Both married second wives named Betty.
- They named their firstborn sons James Alan and James Allan.
- Both had dogs named Toy.

- Their smoking and drinking patterns were almost identical.

Coincidence? Genetic predetermination? Odds are that it is a little of each. In subsequent studies of other pairs Bouchard found the same kinds of similarities: two British sisters who named their firstborn sons Richard Andrew and Andrew Richard, their daughters Catherine Louise and Karen Louise (Karen's name would have been Katherine if family pressures had not intervened); sisters who both came to his laboratory wearing seven rings and a bracelet on one hand, two bracelets and a watch on the other; sisters who, when they were reunited briefly as children, had worn their favorite (identical) dresses. If these were coincidences, they seemed to strike identical twins more often than the general population. Noted Bouchard: "I was expecting to find all kinds of differences because of their different backgrounds, but what leaped out at us were the striking similarities between the twins. I wasn't prepared for it. Nothing prepares you for it."

The Jim twins also gave the physicians in the study something to think about. Both had hemorrhoids. Both had high blood pressure. Both had had two episodes of what they thought were heart attacks. Both had inexplicably gained ten pounds at about the same point in their lives; both had, just as curiously, stopped gaining weight at the same time. And both had complained about a series of on-and-off headaches since the age of 18, a problem that had led to the suspicion that each was a hypochondriac. The headaches were of a type called "mixed headache syndrome," a tension headache that evolves into a migraine. The two suffered the same degree of pain and disability, had the headaches with the same frequency, and described them in precisely the same way. A chance occurrence? A few years ago, physicians believed that mixed headache syndrome was caused strictly by the environment. Because of the Jim twins they are changing their minds.

Flushed with success, Bouchard and his team of 17 psychologists and physicians decided to go after other twins. They devised an experimental scheme one week long that would be used for all pairs. In the psychological portion of the testing, the twins would undergo stress tests, a sexual history, psychological interviewing, psychomotor tests, and a battery of IQ tests that, all told, added up to 15,000 questions.

The medical part of the examination included electrocardiograms, electroencephalograms, pulmonary analysis, blood tests (for HLA antigens and ABO blood groups), allergy testing, eye and dental check-ups, neurologic tests, and a complete physical. The entire procedure would take about 46 hours and cost between $5,000 and $7,000 per pair.

When the Jim twins had gone, the Minnesota team began its search for new pairs. They were in luck. The Second World War had orphaned and separated sets of identical twins. And the rise of sexual freedom, combined with the social disgrace of bearing children out of wedlock, had pressured unwed mothers to put several more such pairs up for adoption. Within 18 months, they had tested another 16 pairs, with about 30 more on tap. In addition, they had found several sets of fraternal twins reared apart and had begun to test them for comparison.

At least one of their new sets offered a treasure trove of information. For while most pairs had grown up in relatively similar environmental surroundings, Oskar Stöhr and Jack Yufe had been raised in fashions that could not have been more opposite. Oskar and Jack were born in Trinidad in 1932 to a Jewish father and a German mother. Soon after, their parents underwent a bitter divorce. Oskar went with his mother to live in Germany. He was raised a Catholic and was on the verge of joining the Nazi Youth movement as World War II drew to a close. He is now a supervisor in a German factory.

Jack stayed with his father in Trinidad. He grew up a Jew. At the age of 17 he went to Israel, where he spent five years working on a kibbutz. On his way home, he stopped off in Germany to find his twin. The first thing his brother asked him to do, through a translator, was to avoid mentioning their Jewish heritage to Oskar's neo-Nazi stepfather. Not surprisingly, the meeting went poorly. Jack returned to the United States, where he now runs a clothing store in California.

The twins were raised by parents of different sexes, with diametrically opposing viewpoints. One endured the devastation of defeat in war, the other the oppressive atmosphere of a country under siege. Neither spoke more than a few words of the other's language nor accepted the other's beliefs. Here was a pair that should have produced some significant differences.

Jack read about the Minnesota study and contacted Bouchard;

Oskar was flown to the Unites States to participate. The two men met for the second time in 47 years at the Minneapolis airport. Oskar wore a blue shirt with epaulets. So did Jack. Oskar sported a well-trimmed moustache. So did Jack. Oskar was wearing wire-rimmed glasses. So was Jack.

The similarities did not end there. As the testing continued the same kind of pattern that had developed for the Jim twins unfolded. Both Jack and Oskar like spicy foods and sweet liqueurs. Both flush the toilet before and after using it. Both enjoy sneezing to startle people. Both store rubber bands on their wrists, dip buttered toast into their coffee, and read magazines back to front.

Because Oskar spoke no English, he could not take all the tests that Jack did. But the results of those that they could take together proved remarkably alike. The translator noted that Oskar's speech patterns in German were the same as Jack's in English. Even Bouchard was moved to note how alike they were in "temperament, tempo, the way they do things." Finally, as the Minnesota team pointed out, the fact that one was raised by a woman and the other by a man seemed to make little difference in their mature personalities. They seemed to offer powerful evidence against the hypothesis that children's personalities are partly shaped by the sex of the person who raises them.

The Jim twins, Oskar and Jack, and the other 15 pairs have left Bouchard and company with a kaleidoscope of tantalizing leads, enough to last many such teams a lifetime:

- One of the areas that showed the least coincidence among the twins was smoking. About half the pairs include one twin who smokes and one who does not. But at the same time, the study seems to provide one interesting bit of evidence pointing to a genetic basis for disease caused by smoking: In at least one set of twins, a heavy lifetime smoker did as well in the pulmonary exam and heart stress test as the non-smoker, implying that both may carry some genetic resistance to smoking's long-range effects on the lungs and heart.

- In those cases where one twin wore glasses and the other didn't, the second twin invariably needed the same amount of correction as the first. Jim Lewis, who had been without glasses at the brothers' first meeting, had indeed worn glasses before, and with the same prescription as his brother.

The tendency for twins to talk and even think alike has been borne out in the tests. Several pairs switched almost immediately into a way of speaking in which one would begin a sentence and the other would end it, even while a third person in the conversation had not yet picked up the drift of what was about to be said.

Bouchard calls any attempts to draw firm conclusions before all the returns are in on the study "little more than gossip." But he agrees that the preliminary findings point to a massive contribution by genes to behavior. The twins' EEGs, EKGs, and brain wave patterns were often so alike that they could be superimposed, one on the other, with only a few variations in the general theme. And their scores on psychological and intelligence tests were usually so close that they varied less than the differences psychologists expect when one person takes the tests twice. Even areas that have always been considered the province of the environment, like specific phobias, seem to have a strong genetic component. One set of twins, raised in distinctly dissimilar home environments (one in a strict disciplinarian setting, the other in a warm and loving household), exhibited the same sorts of neuroses and tendencies toward hypochondria. Another pair turned out to be claustrophobic—both balked at having to enter a tiny, soundproof chamber for one of the tests. And both twins in a third pair were terrified of the water; each managed to "solve" her problem at the beach in the same way: by backing into the surf until the water reached her knees.

The Minnesota team is quick to point out that the material it has gathered is not always exactly what it seems. It would, for instance, be ludicrous for anyone to surmise from the similar names that crop up in several pairs' lives that preference for names is genetic, or from the sisters who each wore seven rings on the same hand that there exists a gene that predisposes one to wearing a set amount of jewelry. Rather, if they are more than coincidence, both characteristics are probably manifestations of other sorts of genetically influenced trends. The women with the rings both had exceptionally beautiful hands with long fingers; the psychologists theorize that both might be predisposed toward accentuating their most attractive characteristic, which would explain the similarity. And for extraordinary correlation in the names? Research has shown that while language (and culture) are traits that have very little to do with genetic predisposition, preference for certain sounds does have a genetic component. With their genetically identical likes and dislikes, the Jim twins and others may be predisposed to pay more attention to someone with a particular sound in her name.

The combination of psychological and physical evidence that they have gathered has astounded Bouchard and his colleagues. It prompted them to note in one of their preliminary papers on the study: "Looking across the twins' educational and work histories and marriages, as well as their psychological responsiveness in the various assessment settings, we find overwhelming patterns of similarities. Having been familiar with the literature on the heritability of temperament, we were not ready for what we found. Worse yet, we do not feel we have adequately captured the phenomenon. Many differences between the twins are variations on a theme more than anything else."

The studies at the University of Minnesota are not the only ones on genes and behavior being undertaken these days. Several others are examining the same relationship from different perspectives. One study, performed by Sandra Scarr also of the University of Minnesota delved into the similarities and differences in social tendencies between parents and children in two groups: one in which the children were adopted, the other in which the children had been raised by their biological parents. In socioeconomic status, ages, and occupations, the families were very much alike.

Scarr's study provides powerful evidence about the influence of heredity on some aspects of people's social tendencies. Among other things, she found almost no consistent similarities between adopted children and their adoptive parents, even after they had lived together for nearly two decades. Children did resemble their biological parents, but only enough to explain a small part of what contributes to sociability. The strongest correlation, it turned out, was between children and their biological brothers and sisters, which is an indication that peer relationships play an important role in our development of social attitudes and responses. It seems likely that children are more apt to copy the social tendencies of their peers than those of their parents, and that the genes play a small but significant role in directing those tendencies.

Another researcher, Ronald Wilson, of the University of Kentucky, has been studying the effects of genes on early development in children by comparing fraternal and identical twins. In one study, he tested 261 pairs of twins at six regular intervals during their first two years of life to chart the evolution of their mental and motor skills. Wilson discovered that development in young children is utterly unpredictable, that their skills increase in no discernible order, advancing sometimes by leaps and bounds in one area while remaining constant in another, then suddenly reversing the trend. In addition, the rates of change were not uniform; a child's progress could differ significantly from one age to the next. Nevertheless, identical twins progressed at almost the same rate, while no such correlation existed among fraternal twins. Children with the same genetic blueprint followed the same course of development; children with different genes did not. The trend was so pronounced that Wilson could predict the state of development of one twin at a particular age simply by examining the other, even though he could not predict future development for either twin alone.

Wilson's second study involved 350 pairs of twins from four to six years old. He tested them for IQ and found that the correlation between identical twins was almost double that between fraternal twins. From this testing he concluded that "each home environment adds its own distinctive impress to the child's cognitive functions, but these influences act as modulators rather than primary determinants." His results indicate that as long as environmental conditions fall within acceptable limits, it is the genetic blueprint that determines the course of infant development.

Several other studies of children's development have shown equally dramatic evidence of the genes' contributions. In the early 1950s, Jean Piaget, the father of child psychology, proposed that a child begins to think symbolically at about the age of two, drawing on memories for the first time to make sense out of the environment. Others have supported his theory, pointing out that the switch from a brain that supports primitive motor and sensory functions to one that is capable of conscious thought is obvious and is accompanied by an extraordinary spurt in the size of the brain itself, which nearly doubles in size between the ages of one and two, with most of that growth coming in the regions dedicated to language memory, attention span, skilled movements, spatial understanding and the ability to plan ahead. Perhaps the "terrible twos" are a function of this change: the beginnings of a child's attempts to cope with a new perception of the world.

Wilson himself has placed another interpretation on his material. He believes we must begin to accept the possibility that mental development is as closely linked to the genes as physical development; that we, as a species, are programmed to develop our mental capabilities at specific stages during our early lives. His hypothesis adds fuel to the growing popularity of the idea that we are governed in part by a complex, finely tuned genetic clock, one that triggers not only such broad physical changes as growth spurts, puberty, menopause, and aging but that also directs the general course of finer details in our lives, that influences the timing of nearly everything biological. The Jim twins' surprising ten-pound weight gain and their carbon-copy headaches, the consistent appearance of certain cancers (Wilm's tumor of the kidney in children, multiple myeloma of the blood in those over 40) in specific age groups, the pattern of aging in cells, the nature of mental development in children, all point to a system that is designed to trigger specific responses at particular points in our lives, as long as critical environmental factors are also present.

Whether we will ever gain control of that clock is still pure conjecture. The delicate balance of gene and environment, the awesome complexity of so many of our traits, the difficulty we have in discovering how tinkering in one area of the gene-environment relationship may affect another, all make it impossible to predict the extent to which we can adjust the clock's timing. But if we can, our tools will be the same as those we are already using to predict the onset of purely physical diseases: the presence or absence of the products of genes, the nuts-and-bolts of genetic prophecy. For any particular characteristic, the more we learn about its genetic components, the more likely we are to be able to isolate the right environmental triggers and to modify them according to the individual's needs.

The paths are the same: Already researchers are searching for the basic molecular causes of various kinds of behavior, the genes that influence them, and the markers that can tell us whether they will occur.

MARKERS AND THE MIND

Men at some time are masters of their
fates: The fault, dear Brutus, is not
in our stars, but in ourselves . . .

Julius Caesar, WILLIAM SHAKESPEARE

Scientists burrowing into the connection between genes and psychiatric diseases are in the midst of a mad love affair with Denmark. Purely by chance, Danish social planning has developed a system that offers an opportunity for some of the most revealing surveys on psychiatric problems ever conducted.

Denmark is valuable not only because its small population is relatively homogeneous and accessible; equally crucial is the Danish tendency (some would call it "compulsion") for keeping remarkably complete and accurate records. The Danes maintain scrupulous lists in three critical departments: a population register that includes the name, date of birth, and address of anybody who has lived in Denmark, even for as short a time as a few months; a national psychiatric register that lists the names and diagnoses of over 90 percent of the Danes who have sought professional help or stayed in a psychiatric institution, no matter what the reason; and, most important, a comprehensive record of every legal adoption that has taken place in the country in modern times, with the names of the adoptive parents, the biological mother, even the presumed father who, in Denmark, must participate in the adoption proceedings.

Records like these have formed the basis for an extraordinary study of the genetic and environmental elements in abnormal behavior, a study that began in 1963 and is still bearing fruit. The researchers who originally conceived it—a team organized under Seymour Kety, a leader in psychiatric genetic research from Harvard Medical School—recognized that the Danish records contained all the elements necessary for an investigation of people who had grown up with two separate pairs of parents: a biological set who had contributed very little to the environment in which the children had been raised; and an "environmental" (adoptive) set, who had contributed nothing to their adopted children's genetic heritages.

The Kety team searched through the registries and compiled an enormous sample of potential cases. In Copenhagen alone, they found 5,500 people over the age of thirty who had been legally adopted by people who were not their biological relatives; in the rest of Denmark, they corralled another 9,000, for a total of 14,500 people who had lived at least 25 years in an adoptive, rather than a biologically determined, environment.

The team had originally come to Denmark with a specific goal in mind—to discover underlying factors in the mental disorder schizophrenia. They were well aware of the difficulties they faced: Schizophrenia has proved about as easy to define as a lovely day. The problem does not lie in getting people to agree on the aesthetic value of, say, the first spring thaw after a long winter. It is more a matter of the borderline situations—the clouds, breezes, slight chills—that make general agreement next to impossible. That is when diagnosing schizophrenia becomes an art rather than a science, a conclusion based at least partly on the subjective impressions of the examiner. For while psychiatrists can generally agree that someone who claims that Martians are communicating with him telepathically (and then acts on that claim) is schizophrenic, they have reached no consensus on the specific constellation of factors—the major disturbances of thought, emotional expression, speech, movement, and behavior—that either confirm or deny the existence of the psychosis.

To begin their study, the Kety team used definitions common to psychiatry in the United States. They agreed to look for three fairly distinct types of schizophrenia: chronic, acute, and borderline. Then they went through the Danish records, matching the adoption lists with the psychiatric registry, and emerged with 34 schizophrenics out of the 5,500 adoptees in Copenhagen, a number not far out of line with the incidence of 1 to 3 percent that appears in the general population. Because two of the 34 were identical twins (with the same biological and adoptive parents), the team now had 33 index cases as a basis for comparison. To complete the experimental design, they chose a control group of 33 normal adoptees who resembled the schizophrenics as closely as possible in such elements as age, sex, and social class.

The scientists now had four sets of relatives to investigate— the biological parents and siblings for each group, as well as the adoptive parents and siblings—for a total of 512 people. And they knew that schizophrenia tends to run in families. By discovering which groups of relatives produced a higher incidence of schizophrenia, they could determine which factor, the genes or the environment, plays a more critical role in the disease. For if schizophrenia is an expression of genetic factors, it would appear significantly more often in the biological families; and if it is an offshoot of environmental factors, it would be concentrated in the adoptive families.

Almost immediately they ran into trouble: there were too few hospitalized schizophrenics among all the relatives to allow them to discern any pattern at all. So the team made a crucial decision to include those whom psychiatrists term *uncertain schizophrenics* and *schizophrenic inadequate personalities,* people who exhibit schizophrenic-like behaviors too mild to be considered strictly schizophrenic. The change was a difficult one to make. Denmark accepts only the classical definitions of schizophrenia; and so the team had to make these diagnoses of so-called spectrum schizophrenia on their own, using only the elaborate records provided by the psychiatric registry.

To preserve the integrity of the investigation, the team created a code for the index cases (the original schizophrenics), the control cases (the normal adoptees), and all the relatives. Nobody who knew the code performed the investigative work.

During their first study, the team managed to survey 463 out of the 512 relatives, and found 21 that fit into the schizophrenic spectrum. When they broke the code that protected the relatives' identities, they discovered this breakdown in the numbers:

Number and Percent of Schizophrenics Among:

	Biological Relatives	Adoptive Relatives
Index (schizophrenic) cases	13/150 (8.6%)	2/74 (2.7%)
Control (normal) cases	3/156 (1.9%)	3/83 (3.6%)

The results clearly indicated that schizophrenic disorders were concentrated among the biological relatives of schizophrenics. But the team was not satisfied; the inclusion of the fuzzier definitions of the schizophrenic spectrum had made them cautious. To test their findings, they decided to have a Danish psychiatrist, Bjorn Jacobsen, interview every relative he could find.

Jacobsen spent two years interviewing relatives who were willing, about 90 percent of those who were still alive. He also received some useful information from many of those who declined to participate, during the course of some long and complicated refusals that he elicited when he asked them why they would not help. Interestingly, even those who died provided some clues about the accuracy of the study. Although they were not included in the final results, most of the deaths were among biological relatives of the schizophrenics; and many were a result of suicides, often a sign of mental disturbance.

Jacobsen worked as the Kety team had, not knowing which relatives belonged to which group. Yet his results mirrored those of the original team. And a second group of psychiatrists that was asked to analyze Jacobsen's results agreed with his findings.

Kety's team used the same analysis on adoptees in the rest of Denmark. The ratios among the four sets of parents were carbon copies of those in the Copenhagen study. Curiously, however, the rate of schizophrenia in all categories was about half of what it was for those living in the city, indicating the importance of environmental factors in schizophrenic disturbances.

Kety has pointed out that his results imply that "there must be two forms of schizophrenia; that which is primarily genetic and that which is primarily environmental." The fact that about half the schizophrenic adoptees had biological families with no signs of the disorder seems to support his conclusion. Speculation about the environmental factors has ranged from the notion that a virus may be involved to the recognition that nearly all the "environmental schizophrenics" were born in winter, suggesting that the risk of birth traumas might be higher during that season. Still other researchers have pointed out that just because schizophrenia didn't appear in the biological families before doesn't mean that genetic components didn't exist; they simply may not have been expressed in the parents. And others have noted that if a virus is involved, as Kety suspects, researchers may ultimately discover a genetic marker for susceptibility to it, another case of an "environmental" cause with a genetic component.

The Denmark studies both confirm and are confirmed by other surveys of schizophrenia. Leonard Heston, a member of the group at the University of Minnesota that is looking into the phenomenon of identical twins raised apart, has found a strong correlation between schizophrenic mothers and the problems of the children they give up for adoption at or near birth. And some twenty studies of identical and fraternal twins that have been conducted since 1928 indicate, on the average, that if one identical twin becomes schizophrenic, the other has about a 50 percent chance of becoming ill as well. The rate for fraternal twins is much higher than that of the general population; they do, after all, share about half their genes and much of their environment. But it is still about a quarter of the rate for twins with identical sets of genes.

The groundwork that Kety's team laid in Denmark has spanned other studies as well. Surveys of adoptees with manic depression have shown that those with a familial history of the disorder have about three times as great a chance of becoming depressed as the general population. And of the 18 suicides that took place among the relatives of both depressives and controls in the Kety study, 15 belonged among the biological relatives of the depressed group, implying a strong likelihood that those exposed to certain stresses and environments may actually be genetically predisposed to killing themselves.

The Kety study does not presume to isolate all the underlying factors that certainly contribute to these complicated behavioral disorders. But it offers researchers a simple, clear-cut design for experimentation that is free from many of the internal biases that plague other such studies. Its capacity to reveal previously unsuspected genetic links to certain behaviors and to confirm the existence of these connections in others make it a powerful tool for future research, an important step on the path to behavioral markers.

Markers and Alcoholism

The links between genes and our response to alcohol appear in all sorts of genetically distinct groups. When Japanese, Taiwanese, and Koreans drink amounts of alcohol that have no visible effect on Caucasians, their faces flush markedly and they exhibit mild to moderate symptoms of intoxication, a difference that may be tested early in life and has been found to have no relation to other activities. Studies of identical twins have revealed that if one twin is an alcoholic, the other has a 55 percent chance of becoming one as well, while the level of concordance among fraternal twins is only 28 percent. The survey of alcoholics among Kety's Danish adoptees found that the men had about four times the risk of developing alcoholism if one of their biological parents was alcoholic. And Swedish studies on the relative influences of the environment and genetics have found that heritability may be as high as 90 percent for some forms of the disease.

These statistics do not mean that any one child of an alcoholic will also become an alcoholic. The 55 percent concordance among identical twins and the fact that women are far less likely to become alcoholic than men—possibly because they may have less contact with heavy drinkers—should give some indication of the importance of environmental factors in the disease. Nevertheless, work is now being undertaken to determine exactly what the genetic components of alcoholism are.

Just as "cancer" is a term that encompasses dozens of different specific problems, alcoholism may be caused by any one of several mechanisms. Among those identified so far as possible culprits are the rate of intoxication, the rate of elimination of ethanol (pure alcohol) from the blood, addictability, and the susceptibility of the liver, pancreas, brain, and fetal tissues to the complications that chronic alcoholism can cause. These factors, either alone or in combination, may underlie any one individual's propensity toward the disease.

With an eye toward these difficulties, researchers have begun to test different people for possible inherent differences in their reaction to alcohol. One such study, performed by Marc Schuckit and Vidamantas Rayses, two psychiatrists at the University of Washington, has isolated a possible marker for some forms of the disease.

Schuckit and Rayses surveyed 304 healthy males from the university and selected 20 who had an alcoholic parent or sibling. Then they matched them to a control group with no alcoholic background, but with similarities in age, sex, marital status, and drinking history. The two groups were given doses of ethanol that varied according to their body weights. Then samples of their blood were tested for the appearance of acetaldehyde, a product of the body's efforts to break down ethanol and a poisonous substance implicated in some of the more debilitating effects of alcoholism. A clear difference appeared between the two groups: Those with family histories of alcoholism had almost twice the amount of acetaldehyde in their blood as the control group.

As the two researchers have noted, these initial findings have profound implications. An increased concentration of acetaldehyde might actually change the way alcohol makes susceptible people feel while they are drinking and may provide a possible physical link to the psychological aspects of addiction. The genetically caused higher concentrations may also make those predisposed to alcoholism more vulnerable to internal damage from the by-products of alcohol metabolism.

An even stronger connection has been discovered between alcoholism and psychiatric disturbances. Some alcoholics develop severe memory loss as well as thought disorders. Frequently, psychiatrists describe these people as being in an "alcoholic deteriorated state," a blanket generalization that obscures the fact that we don't really know the cause of the disorders. But some people, mostly of European extraction, combine these psychiatric symptoms with eye and balance disturbances. When physicians discovered that this combination could be distinguished from the more vague "alcoholic deteriorated state"—and was concentrated in a single population—they gave it a name (Wernicke-Korsakoff syndrome) and began to search for a genetic cause.

Soon they found one. They noticed that the nerve cells in people with Wernicke-Korsakoff syndrome do not function properly and linked the problem to a defective enzyme, transketolase.

People with defective transketolase need more thiamine, a B vitamin, than the normal population. Fortunately, the average diet supplies enough to protect them. Alcoholics, on the other hand, have notoriously poor nutritional habits. It is not that they simply are more susceptible to Wernicke-Korsakoff, but that they also do not ingest enough Vitamin B to protect themselves against the defective enzyme.

Some physicians have suggested that Wernicke-Korsakoff syndrome would be alleviated if liquor were fortified with thiamine. Others have mentioned the possibility of screening alcoholics for the defect so that, if nothing else, particular attention can be paid to their diets.

Disorders such as alcoholism and schizophrenia are generally far less amenable to genetic prophecy than other diseases. They are difficult, if not impossible, to define; they are usually caused by the interaction between an entire constellation of genes and the environment; and the environmental factors themselves may prove to be impossible to pin down. Nevertheless, our use of genetic tools to address these issues is getting results. Our ability to uncover the underlying mechanisms of these diseases improves almost daily. As we move toward an understanding of how behavior works, we come closer to the time when we can define these problems through both their environmental and their genetic components; ultimately, we may reduce what we don't know to the point where, in a practical sense, it no longer matters.

The Biology of Behavior

The brain has long been one of the most misunderstood organs in the body. Unlike the lungs, the kidneys, or the heart, the brain has no obvious function. It appears at first to be an amorphous mass, a three-pound lump of grey matter without distinguishing characteristics, with no clearly defined connections to the normal patterns of flow between the body and the environment. The brain is the hub for an enormous plexus of nerves that extend to every organ, muscle, and sensory surface of the body, but its role is far from clear. As a result, ancient scientists put little faith in its importance. The Sumerians and Assyrians believed that the soul resided in the liver. Aristotle campaigned for the heart, relegating the brain to a secondary role as a thermostat which regulated the cooling of the blood during its travels.

Gradually, however, the brain began to receive the attention it deserved. Scientists began to realize that its simple, undifferentiated appearance masked an organ far more complicated than the kidney or heart. They began to probe its function, finding that electrical shocks could cause frogs' legs to jump. They learned that it was filled with tiny electrical circuits. They looked through microscopes and found that it was composed of various kinds of cells. By the middle of the nineteenth century, scientists felt comfortable enough with their knowledge of the brain to compare its functions to the tangled interconnections of the telegraph, long before they had access to the facts that proved it to be so.

Now we know what the brain is and does. We recognize that it contains the maze of the mind, that it consists of a highly organized assemblage of cells that receives, synthesizes, and responds to information, and that it can actually initiate action on its own. And we compare it as best we can to that epitome of modern technology, the computer, having discarded the idea of the telegraph as too simple to encompass all the brain's workings. Nevertheless, while we have clarified some of the brain's mechanisms and have learned how some of its parts are organized, we still have almost no idea of what it all means. We can't even guess, for instance, what the mind is—if it is, indeed, any *thing* at all—how the centers of behavior are structured, how that vast constellation of nerve cells, or neurons, keeps it all under control. And we are a long way from determining precisely how the brain responds to the environment: what parts it helps us control, and what parts strongly affect it.

Forty years ago, the possibility that the environment might influence the brain was never even raised. Scientists knew that the central nervous system was bombarded with stimuli, faced with the task of making decisions based on a crush of information relayed by its sensory scouts in the muscles, skin, eyes, tongue, ears, nose, and internal organs. But they assumed that the brain functioned in splendid isolation. They believed that it was so critical to our existence, so delicately tuned, so devoid of back-up mechanisms, that the rest of the body acted as a buffer between it and biochemical elements in the environment that might upset its sensitive workings. Some hypothesized a chemical barrier that prevented environmental insults from reaching the brain at all, that preserved the brain as a command post of intelligence in the same way the Rocky Mountains now protect the computerized brain center of the Armed Forces in the event of a nuclear attack. Only strong environmental intrusions—a billy club on the skull, perhaps, or a dose of alcohol—could reach it.

Others, however, disagreed. They argued that the brain would ultimately prove to be susceptible to environmental influences. Sigmund Freud himself predicted that the types of behavior he was studying would eventually be discovered to have biochemical foundations. But only in the past few decades has the medical community developed the tools to explore this possibility: and only in the past few years have the established centers of psychology accepted the fact that various forms of behavior have biochemical, physical components.

The investigation of the nature of diseases passed through three stages of sophistication during the past few centuries, as scientists first viewed disease as utterly external, then began to explore its microscopic components, and finally formulated theories based on these revelatory discoveries. Research into the brain can be broken down into the same groups. Today we are still in the second stage of exploration, far from an understanding of the more complex mechanisms (memory, consciousness, intelligence, or learning), but gathering insights about the pathways of information. And we are beginning to understand how the brain works on its most elementary level: that of transmitting information among individual nerve cells, or neurons.

Imagine that every one of the nearly five billion people in the world has several dozen telephones at his or her disposal. Only a few of these phones can call long distance, but a series of local calls can eventually transfer a message from any one point to any other. Some people use all their phones; others do not. But the *potential* for calls to be placed to and from every single phone exists at all times. That, in its simplest form, describes the almost inconceivable complexity of the interaction between the brain's countless (estimates range from 10 billion to a trillion) neurons. A single neuron has been known to support as many as 200,000 junctions with other nerve cells. It can be besieged by a medley of different signals from hundreds of them at any one time. And out of this barrage, it manages to extract a single piece of information, synthesizing it from among all the incoming signals, translating it to a simple electrical impulse, and passing it on to the next neuron.

Neurons transmit information both electrically and chemically. Messages moving within a single neuron are electrical in nature; but when they move between neurons, they are transformed into a flood of chemicals which are released by the transmitting neuron. Some of the chemicals are tiny, tangled fragments of protein, called neurotransmitters. They float toward the next nerve cell and attach themselves to it, fitting like keys into the locks of special receptors in the neuron's tip. Their presence can trigger or suppress an electrical impulse in the cell, and the message is moved onward from neuron to neuron, guided by the alternating chemical and electrical pulses through the labyrinth of the brain.

The different neurotransmitters are already famous in their own right. There are the *endorphins*, morphinelike substances that can smother the sensation of pain; *acetylcholine*, the most abundant transmitter, found in nearly every neuron; *dopamine*, implicated as one of the critical transmitters for behavior and learning; *serotonin*, which seems to be linked to sleep and depression; *norepinephrine*, serotonin's alter ego, which has been tied to arousal, aggression, and heightened physical activity; and many others. Some neurotransmitters are chemically similar to hallucinogens: the structures of norepinephrine and mescaline (the active substance in peyote) are very much alike; and serotonin and the street drug DMT—a compound like LSD that often causes psychotic reactions—are also closely related. One popular but unproven theory is that we normally produce our own internal supply of hallucinogens, and we actually do produce enzymes that are *capable* of their production. We may someday discover genetic defects that cause some of us to overproduce certain neurotransmitters in response to an environmental stimulus like stress.

The specific roles of neurotransmitters in behavior depend on two factors: Where they are located in the brain—that is, which group of neurons (memory, vision, muscular action) they influence; and whether they *excite* neurons (turning them on and triggering electrical impulses) or *inhibit* them (preventing them from firing). Neurotransmitters generally either excite or inhibit neurons, not both. But their impact on behavior depends on the nature of the group of neurons they are affecting. A lack of dopamine, for instance, may cause Parkinson's disease in one area of the brain, emotional disturbances in another, and a disruption of hormone regulation in a third.

Current research indicates that the most common forms of mental illness (manic depression, depression, and schizophrenia) are all significantly affected by the levels of one or more of these transmitters in different parts of the brain. The use of psychoactive drugs confirms these findings: drugs that have powerful effects on behavior often act by modifying the levels of neurotransmitters.

The existence of neurotransmitters and the effectiveness of psychoactive drugs illustrates the biochemical nature of behavior. But how does heredity influence this relationship? It seems that the genes play two basic roles in the character and function of the brain: Genes form the blueprint from which the protein building-blocks of the brain are derived. As such, they are crucial in helping to determine the nature of the brain's organization—how, in each individual, the centers of various kinds of complex patterns of behavior are arranged. Researchers now believe that the genes may provide an excess of nerve cells—far more than the brain actually requires. The environment then selects from among the available pathways to shape a working brain, determining which neurons are connected to which. Among the critical questions still unanswered: What is the plan that determines how genetic instructions are translated into the brain's wiring? How does the environment manage to direct the pathways of neural connections? How much of a single neuron's affinity for other neurons comes from the hereditary blueprint and how much is left malleable?

Second, genes also help determine the concentrations of neurotransmitters and other chemicals that are available in specific areas of the brain. The genetic connection has been borne out by several studies involving the neurochemistry of mice. The studies show that the influence of genes varies; the levels of some neurotransmitters are probably mediated by a single gene—implying that the genetic contribution is relatively straightforward—while others are undoubtedly directed in part by groups of genes. One recent study has confirmed the role of genes in determining the number of dopamine-producing neurons that grow in the brains of mice, which is an indication of how genes might control chemical production.

Genes may also control the level of the various enzymes that are involved in the transmitter-receptor connection. One of these enzymes, called monoaminoxidase (MAO), seems to help receptors rid themselves of transmitters that have done their job. By removing transmitters after a certain period of time, MAO blocks further transmission of a signal until a new flood of transmitters is loosed. Low levels of MAO, which would allow neurons to be in a constant state of excitement or inhibition, have been tied to chronic levels of schizophrenia and manic-depression; in one such study, low MAO activity was linked significantly to higher levels of psychopathology in otherwise healthy volunteers. Since then, tests of MAO activity had been used successfully in experiments to predict the likelihood that some college students would have psychiatric difficulties. Several tests have disclosed a suicide rate among men with low MAO activity and their families that was eight times higher than that of men with high MAO activity. Other studies contradict these findings, and scientists generally feel that until MAO's role is better understood, it will be useful only in experimental settings. Nevertheless, the roles of enzymes in behavior are slowly being recognized. Ultimately, they might become more important to the science of predicting susceptibility to mental disorders than the neurotransmitters themselves.

Research like this into the specific influences of genes on the most fundamental relationships in the brain have set the stage for further work. Thus far, the findings have been tantalizing, but incomplete. Some pathways have been proposed, but they are complicated; many mechanisms are still unknown. Nevertheless, even from so little information, a broad pattern is emerging. An astonishing parallel seems to exist between the way genes and environment interact in behavior and the way they work together in disease. Again, the genes provide a rough template, a blueprint of general organization and structure, a set of potential links with one's behavioral ancestry. The environment does the fine-tuning, choosing from among these possible paths, whittling the broad outlines into a polished whole. The relative influence of genes and environment may change according to the characteristic:

breathing, for instance, is under tight genetic control, while our more complex behaviors are far more subject to subtle environmental pressures. But the general theme—of genes, environment, and their interaction—remains the same.

As if to prove the validity of this idea, the parallels do not simply stop at the general level of behavior. Even in the biology of the brain, the same relationship exists. The brain's own organization provides the "blueprint" for behavior, defining the possibilities; and the neurochemicals, the transmitters and enzymes, constitute the "environment" which selects from among the blueprint's alternatives. The pattern is causing researchers to delve deeply into the brain's molecular structure for clues to how it works.

The Making of a Marker

In the mid-1970s, David Comings of the City of Hope National Medical Center in Duarte, California, began a search for mutant, or altered, proteins in brain tissue. Comings was especially interested in the brains of patients who had succumbed to Huntington's disease, an insidious, genetically transmitted disorder that causes rapid emotional and intellectual deterioration, usually around the age of 40. Huntington's disease has struck down, among others, Woody Guthrie, the folk balladeer who was at first misdiagnosed as schizophrenic, as were many others with the disease.

Huntington's disease is triggered by a dominant gene; only one copy of the gene, from one parent, need be inherited for the disease to occur. Although scientists have not yet located the gene, their statistics prove its dominance; on the average, one out of every two children of a parent with Huntington's disease also becomes ill.

Because of the nature of the disease, it is also clear that the gene for Huntington's disease produces a protein that acts on the brain. Because of these characteristics, Comings felt certain that the brains of Huntington's victims would contain a marker, and that, if he could isolate it, he could predict who would eventually come down with the disease.

Coming's experiment followed a classical method of isolating proteins in tissue. First he took slices of matter from the brains of both Huntington's victims and a control population who had died of cancer, heart disease, and other unrelated causes. Then he minced his samples, mashing them to a smooth pulp, and suspended them in solution. From the solution, he was able to extract a purified sample of the proteins. He spread them on a gel (a viscous substance flattened between two plates of glass) and subjected the concoction to a steady flow of electricity.

Proteins have different weights and electrical charges that can be used to tell them apart. As the electricity passed through the gel, the brain proteins began to migrate in different directions, some attracted to the source of the current, some repelled by it. The distance they moved depended on their size and weight. When Comings turned off the current, his gel had developed a characteristic set of spots at points where identical proteins had gathered. Now he had a map of the brain proteins.

If the experiment is performed correctly, specific proteins always move to the same place on the gel, making them easy to identify. When Comings looked at the pattern created by the brain proteins of Huntington's victims, he noticed a new spot, one that he had never seen before; somehow, he had managed to isolate a previously unknown protein.

Tests on other tissues failed to elicit the protein again; it seemed to crop up only in the brain, an indication that it was produced only when the genes coding for the brain were turned on and for no other tissues. But the protein occurred just as often in the control population as it did in those with Huntington's disease. It was not a marker for that particular disorder.

Still, the protein did exist; and it appeared most often in samples taken from brain sites believed to be involved in psychoses and neurologic diseases. Therefore, Comings changed his experimental goals. He began to search for the appearance of the protein in the brains of people who had had multiple sclerosis, a neurologic disease, and those of suicides who had exhibited signs of depressive illnesses.

Comings examined 276 brains in all—152 controls, 52 from people who had had diseases like multiple sclerosis, and 72 from manic depressives, depressives, schizophrenics, and alcoholics. His findings provided some tantalizing clues about the nature of "mental" illnesses: The protein, which he christened *Pc 1 Duarte,* appeared in all groups, but much more frequently among those in the experimental groups. Of the control group, 2.5 percent were homozygotic (with two copies of the gene that controlled the protein). The group of depressives, schizophrenics, and alcoholics, on the other hand, contained 12 percent homozygotes and 64 percent heterozygotes (with only one copy of the gene); and of those who had had multiple sclerosis over half—55 percent—had at least one copy of the gene.

Our understanding of depression—what it is, what causes it, how it works—is still in its formative stages. But we do know several things.

- Depression is a disturbance of mood. It is not a single, all-encompassing disease, but several smaller ones: Recently, researchers have classified the major categories as bipolar depression (or manic depression), in which an individual may plunge from the extremes of excess energy and hallucinations to the depths of lethargy; and unipolar depression, depression without mania, which is perhaps ten times more frequent than bipolar depression.

- Depression can manifest itself in mild and severe forms. Generally, psychiatrists distinguish between two levels, calling those who are less severely affected "neurotic" and those whose depression is incapacitating "psychotic." Nevertheless, there is very little agreement about where one level ends and the next begins. Perhaps the simplest definition—until specific symptoms can be pinpointed as "neurotic" or "psychotic"—is the old saw that "neurotics build castles in the sky, psychotics live in them, and psychiatrists collect the rent."

- There is a clear distinction between the mild unhappiness that everyone feels once in a while and the chronic, organic depression felt by those suffering from mood disorders. Even so, depression is the most common of all mental conditions; the National Institute of Mental Health has estimated that up to 15 percent of all adult Americans suffer the symptoms of depression in a given year.

As vague as these descriptions may sound, they are advanced and highly technical compared to clinicians' understanding of what causes the condition. Theories have variously pointed to social, psychological, biochemical, and genetic factors as the key to the disorders. Today, at least, the most likely explanation is that a combination of several elements is involved in any single case; until researchers can define organic depression more precisely, that is as comprehensive an explanation as we will have.

Nevertheless, their search will continue to uncover discrete parts of the puzzle, each one taking us closer to understanding the whole. Comings' findings offer powerful argument for the existence of at least one specific genetic factor. If his results are confirmed—if, in other words, the Duarte protein is found to be either a predisposing factor in depression or a link to such a factor—scientists may have one predictive mechanism for the disease.

So far neither the protein nor its gene has been isolated, and Comings does not presume to guess exactly how it functions. He notes the possibility, however, that in the near future patients who complain of depression or heavy drinking could be tested for the Duarte protein, and followed more closely if it appears. Perhaps the protein itself will ultimately provide a better explanation of exactly what the disease is and how it works.

The links between the Duarte protein, multiple sclerosis, and depression are still experimental and statistical. But evidence of their association is appearing in other quarters as well. Although no significant associations have been reported between depression and multiple sclerosis (an association that might be expected if the Duarte protein is a marker for both), one recent study noted that patients with family histories of depression were more likely to carry the antigen HLA-B7 at a ratio of 55 to 19 than those without such a family history. And HLA-B7 is one of several antigens that have been tied to the incidence of multiple sclerosis.

But even if the Duarte protein is found to be linked to depression, the problem arises of how to test for it; a genetic marker found only in the brain is particularly difficult to isolate. So far, the only known way is to take a slice of grey matter and analyze it, a procedure that few people in their right minds would agree to. Until recently, the sheer impracticality of cutting into inaccessible, localized areas of the working brain had researchers stymied. They could test for certain proteins in cadavers and animals, but testing or screening living humans was out of the question.

In the near future, however, that problem may be licked. The stunning advances in biotechnology that have taken place over the past few years have resulted in techniques that give us an opportunity to explore the core of the brain's hereditary blueprint—the genes themselves—without ever intruding on the brain's own sanctuary, the skull.

The conceptual breakthrough that makes this possible comes from a delightful bit of reasoning: Proteins are produced by genes, and every cell in the body (except mature red blood cells) carries the blueprint—the entire genetic code—for every protein the body manufactures. Therefore, the genetic maps for brain proteins like the Duarte protein also exist in each body cell. The genes may be permanently switched off (they may not be needed in the creation of, say, skin cells); but they are there. Now scientists have discovered how to locate them, whether or not they are functioning.

The technique grew out of a series of experiments performed in the late 1960s that discovered how bacteria manage to survive in a hostile world. It turns out that bacteria are not immune to invasion from outside; they are constantly under attack by viruses that use the bacteria's reproductive mechanisms to reproduce themselves. Viruses cannot reproduce without bacteria. So they attach themselves to their prospective hosts and inject their DNA through the cell walls, using a hypodermic-like device they have developed. The viral DNA attaches itself to bacterial DNA; and when the bacteria reproduce, they copy the foreign DNA as well as their own. In many cases, the viral DNA manages to reproduce faster than the bacteria do, the tiny viruses grow and fight for space until the bacteria, stretched beyond endurance, explodes. The new viruses float away, in search of new hosts.

To protect themselves, bacteria have developed a group of bacterial enzymes, called *restriction enzymes*, that act like microscopic assassins, searching out pieces of foreign DNA and snipping them into useless, nonfunctioning lengths. Each restriction enzyme works by recognizing a specific genetic code—a small segment of the DNA message that the bacteria do not normally have. They actually "recognize" DNA that does not belong and destroy it before it can do any harm.

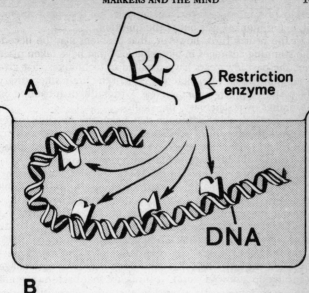

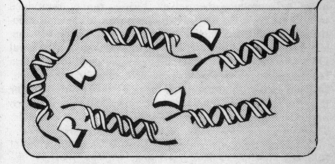

4. Directly identifying genes. (A) Restriction enzymes, which cut DNA only when they come across specific DNA codes, are added to a solution containing DNA. (B) The enzymes migrate to sites containing their codes and slice the DNA. Because the shorter lengths of DNA are measureable, scientists can find out whether a certain gene is present simply by measuring the lengths of DNA—discovering whether a specific sequence exists by determining whether or not it has been cut.

In the late 1960s, Herbert Boyer, a researcher at the University of California, isolated the first of these tiny biological scissors. Since then, dozens more have been identified. Because each restriction enzyme recognizes one, and only one, sequence of DNA, the enzymes are now used routinely in laboratories to chop up lengths of the molecules at sites chosen by scientists. Already, they have been instrumental in the work that led to the first bacterial production of such potentially valuable products as interferon and human insulin.

But the enzymes are only tools. Just like any hammer or electric drill, they can have many different functions. Most recently, they have been used to identify specific genes.

The technique works like this: If a scientist knows what the genetic difference is between a healthy gene and a defective gene, and if he has a restriction enzyme that can cut the defective gene at the point at which its code is different, he can identify the defective gene simply by determining whether or not a sample of DNA can be cut by the enzyme. The healthy gene, with no code that the enzyme can recognize, remains uncut; the defective gene is snipped apart at the site of the defect.

Already scientists have identified the gene for sickle-cell anemia by this method. Now they can actually discover who will contract the disease by testing for the gene that causes it, and not merely by searching for that gene's protein product. The same technique may ultimately be used to identify *any* gene, including those that contain the information for producing brain proteins like the Duarte protein.

Ultimately, these techniques may do far more than simply pinpoint specific genes. If enough restriction enzymes are discovered, or, perhaps manufactured synthetically, we may be able not only to identify large segments of chromosomes but the smallest units of heredity—the base pairs themselves, the rungs on the ladder that Francis Crick and James Watson described some 30 years ago. That astonishing prospect is not nearly as distant as some might think. One prominent scientist in the field, Park Gerald of Children's Hospital Medical Center in Boston, Massachusetts, has remarked: "In the near future we will be able to hand you a list of every one of the three billion base pairs on the human chromosomes. It is no longer a matter of innovation to do this. It is now just a question of time and tedium."

With that kind of precision, genetic prophecy will become

a science of specific molecules. We will no longer have to depend upon the inexact and indirect art of measuring protein levels in the blood for information; we will be able to head straight to the source.

THE BATTLE ROYAL

There is not just one intelligence but
rather several. And I would like to know how a
Nobel Prize winner would feel if he were
by some set of circumstances to find himself alone in a
jungle faced with the need to survive.

FRANÇOIS JACOB

If you feel in the mood for a fight, here is a surefire way to start one: Go to the bar or party nearest you and loudly discuss your views on the relationship between genes and intelligence. No matter which point of view you decide to take, you are guaranteed to find somebody who is willing to go to the wall for the other side.

The battle over what controls the mind is reminiscent of the range wars that enveloped the American West a century ago. The territory is wide open; the frontiers have yet to be fully explored; not even the terms are fully defined. Yet already both sides have staked out their domains. Each has sketched doomsday scenarios if the other side wins. And neither can tolerate neutrality. "If you ain't with us," the saying goes, "You're against us."

Curiously, the problem is not with the scientific evidence itself: that is building in a clear and obvious direction. The difficulties arise when personal beliefs and biases get in the way of interpretation, when politics take the place of logic, when too little information is used to support too expansive a theory.

The protagonists can be divided into two groups, each one a perfect subject for caricature: The environmentalists say: "The genes play an insignificant role in intelligence and personality. They simply cannot be isolated and examined. The power of upbringing and culture is so pervasive that any set of behaviors ascribed to the genes can also be linked to environment. Testing for genes and announcing positive results plays right into the hands of those who are searching for a scientific rationale to bolster the presumed superiority of a race, sex, or geographical group."

The sociobiologists say: "The genes are the power behind the behavioral throne. The human organism is merely DNA's way of making more DNA. Personality and intelligence are genetic vehicles for survival. The environment merely provides a set of variables from among which the genes select. If we refuse to test for genes, we are placing a political agenda in the path of the search for knowledge."

These descriptions may at first seem like oversimplifications of the serious arguments presented by thoughtful adversaries. In many cases, they are. But the arguments battled about in public are seldom serious or thoughtful. On the one hand, there are genetic advocates like William Shockley, the Nobel Prize winner who has agreed to permit his sperm to be used in a scheme of artificial insemination whereby women of "superior" intellect are fertilized by the genes of winners of Nobel Prizes for science (but not for literature or peace). The rationale behind this is obvious. But it violates some of the most basic rules of genetic inheritance. Shockley has several children of his own, all startlingly average. When asked why they are average, Shockley placed the responsibility on the limited intellectual capacity of his first wife.

On the other hand, there are politically motivated groups of scientists like Science for the People, in Cambridge, Massachusetts, which hold that all science and research should be answerable to the public. Some of its members act as devil's advocates for any genetic study that has human implications, no matter how restricted or accurate its conclusions might be. It is not always their fault, of course. Journalists know where to go when they want a fiery quotation to spice up a story. With so many studies being carried out, it is almost impossible for any one person to keep up with them all. So journalists frequently call members of Science for the People, present the facts as they see them, and wait for the thoroughly predictable response. When it comes, everybody is happy. The journalist has gotten his quotation, the socially conscious scientist has free publicity for the cause. Unfortunately, the public has nothing to go on but a few choice words balanced against a mass of carefully derived scientific data. How can people discern which is which?

The bottom line, however, should be solid evidence. And solid evidence demonstrates a clear and growing link between genes and intelligence. The connection is not always denied, of course. Everybody agrees that a child with Down's syndrome (with a third chromosome added to the 21st pair) exhibits the characteristic flattened posterior skull, the widely spaced eyes, the short, stocky build, and, quite often, profound mental retardation. Genetics is also recognized as the critical element in children with PKU—children who can now be protected against retardation by a change in diet. And it is accepted as the cause in the second most common form of retardation for men: the "Fragile X" syndrome, a disorder in which one of the legs of the X chromosome breaks away from the main body and dangles, attached to the chromosome's body only by a thin thread of DNA.

In these instances and others, it is easy to pinpoint "intelligence:" intelligence is what people suffering from these syndromes do not have. They fall short, to some degree, by nearly every yardstick we can imagine: verbal and spatial ability, memory, and reasoning. More importantly, they often cannot fend for themselves in *any* society or environment. And that is a measure that everybody can understand.

The problem, then, is not whether genes can influence intelligence—they do. It is, rather, whether they influence *normal* intelligence, and, if so, how they interact with environmental factors to enhance or limit the elements that, together, comprise an individual's mental capacity. And the question becomes: Can genes create subtle differences within groups and among individuals? Or is the role of training and culture so powerful that it obliterates the genes' fine-tuning, the little predisposition they might provide?

It is here that the discussion breaks down. The possibility that genes might affect levels of intelligence, no matter how subtly, raises the spectre of a world in which children are bred for brains, in which those born with lower IQs are consigned to the bottom levels of society. In less extreme cases, it offers the rationale for major abuses of power, for political systems based on repulsive theories of racial superiority, for insidious incursions into individual freedoms. Genetic "information" has been used to justify the sterilization of blacks in the earlier part of this century, to limit the numbers of certain ethnic groups that wished to immigrate to the United States, and to foster programs of mass extermination in several nations over the years. Few are willing to regard these horrors as aberrations that cannot occur in more enlightened times.

Yet, despite our natural desire to be more careful next time around, the scientific evidence still remains. Psychological tests, studies of fraternal and identical twins (and identical twins raised apart), and investigations within families are remarkably consistent in finding that intelligence is probably the most heritable among normal behavioral and personality traits. Time after time, they indicate that at least 50 percent of intelligence is directed by the genes.

Or rather, that something we choose to call "intelligence" has a genetic foundation. The arguments as to what intelligence includes and how we can measure it are far from over.

What is intelligence? When faced with this question, many psychology textbooks turn coy. "Intelligence is what the intelligence test measures," say several, using the term to define itself, implying that intelligence is whatever the test's interpreter wishes to make it. Dictionaries variously describe it as the "faculty of thought or reason" and the "capacity for reasoning, understanding; aptitude in grasping truths, facts, meaning, etc." All this would be fine if we were talking about some vague concept and not something which has precipitated attempts at genocide.

Unfortunately, science is not yet very good at testing things that are explained by words as frothy as "faculty," "thought," "reason," "understanding." As a result, we really have no operational definition of intelligence. And any assumption that our common use of the word has anything to do with the way a research scientist means it is false.

Intelligence, as we know it, is inextricably bound to culture. Environmental factors even before birth play an enormous role in determining which of the available mental pathways will be used and which will remain fallow. And each society's different emphasis on specific skills—our Western civilization's preference for verbal and analytical abilities, for instance—means that while intelligence may be specified within a culture, the definition cannot be extended to include groups outside of that culture. Trying to apply the rules of Western "intelligence" to an Australian aborigine (who, at least in times past, was concerned with developing mental faculties and perceptions of which we are now only dimly aware) is like trying to hold mercury in your hand. That is why most researchers agree that measurements of intelligence are valid at best only within homogeneous cultures. Comparing the IQs of individuals in two different cultures makes no rational sense in the face of their different values.

This contention recently received support from experimental evidence gathered by Sandra Scarr of the University of Minnesota, who studied the results of IQ tests given to black children adopted into white homes. Normally, blacks score an average of fifteen points lower than whites on the most commonly used IQ tests, a finding that has led to many scientist's speculations about the innate inferiority of blacks. But Scarr found that black children raised in white homes gained an average of 16 points over those remaining in black homes. Furthermore, the younger a child was when he or she was adopted, the greater the difference in IQ. The extreme change in environments had a major effect on the children's scores, which is an indication of the influence on culture on whatever the test was measuring.

Do these findings mean that genetics plays an insignificant role in intelligence? Not at all, for Scarr also found that, within each family, the IQs of biologically related children mirrored each other more closely than the IQs of unrelated children. The fine-tuning that a family's environment provides cannot fully mask the general trends directed by inheritance. Clearly, genes play a role in determining IQ, even though we still do not know exactly what it is that IQ tests measure.

The controversy surrounding IQ tests has forced researchers away from attempting to measure intelligence as a single, unified entity. In fact, most prefer not to speak of "intelligence" at all, opting for the concept of "intelligences"—the separate, distinct, and (some claim) measurable components that, together, comprise mental capacity. They refer to verbal ability, word fluency, perceptual speed, memory, numerical ability, reasoning, and spatial ability, assuming that if they cannot measure "intelligence" accurately, perhaps its components will at least afford them a glimpse of the real thing. Yet even these so-called cognitive abilities are susceptible to vague and contradictory definitions. One researcher, for instance, examined others' studies of spatial ability and found that they had variously measured it as the ability to visualize two dimensions, the ability to visualize three dimensions, and the differences in perception between the left and right hemispheres of the brain. These are three different characteristics that may or may not be related, yet all have been tested to measure the same cognitive trait. Clearly, the confusion about what intelligence actually is goes right to the core of the scientific experiments that are supposed to measure it.

Does this necessarily mean that IQ tests are invalid? Not at all; for they *are* consistent, which means that they are definitely measuring *something*, even if we may not be able to label exactly what it is. IQ tests, for instance, make it possible for scientists to identify the causes of mental retardation simply by analyzing the test results: people with Down's syndrome have test patterns and results that are characteristically different from those who have PKU; and these groups' patterns are strikingly different from groups affected by still other problems.

In addition, genetic markers for intelligence, or, at least, markers that can predict performance on IQ tests, are beginning to surface. So far, at least three have been discovered.

IQ and Fragile X. The "Fragile X" chromosome was discovered in 1969, when Herbert Lubs, a scientist looking into the history of mental retardation in one family, came across the characteristic dangling leg hanging from the X chromosome in one son. He published his results, and others began to look for the disorder elsewhere. Nevertheless, for eight years, the Fragile X syndrome disappeared; nobody could identify another case, and most researchers began to assume that it was a one-time phenomenon.

Then, all of a sudden, the Fragile X surfaced again. A young researcher in Melbourne, Australia, discovered it in eight separate families and established it as an important factor in mental retardation.

Why did the Fragile X disappear for so long, only to reappear in one scientist's studies of a single group? The reason was pure scientific serendipity. The laboratory in which it was isolated had switched its methods of culturing cells from an old, accepted technique to a new one. Fragile X can only be identified under very limited laboratory conditions; the new techniques made it stand out.

Since 1977, cases of Fragile X have popped up everywhere and in every social class. It occurs primarily in males, because males carry only one copy of the X chromosome (their sex chromosomes are designated "XY"); females, who carry two X chromosomes ("XX"), have a normal X that can compensate for, or mask, the Fragile X. Nevertheless, females do carry the Fragile X trait. That X is now thought to be the cause of decreased mental development and mild mental retardation in about 30 percent of the women who carry it.

Scientists exploring the significance of the Fragile X chromosome have theorized that the chromosomal damage may occur at a point which, with less drastic alteration, leads to learning disabilities and dyslexia. While no simple test for the syndrome yet exists, prenatal diagnosis may soon be possible. If so, and if these theories turn out to be correct, the Fragile X chromosome may turn out to be a marker for specific abnormalities in mental development.

The possible uses of such a marker are clear. If we can pinpoint the cause of certain learning disabilities, we may be able to cure, or at least treat, them. On the other hand, if the Fragile X is only peripherally connected to these disabilities (affecting only some of its carriers), we must be extremely careful to avoid stigmatizing those who might otherwise grow up normally.

IQ and PKU. The profound retardation suffered by untreated children with PKU affects those who have picked up two copies of the PKU gene—one from each parent. But what about the carriers? Do those who have only one copy suffer any ill effects?

Recent tests of the IQs of these carriers indicate that they do. On the average, parents of PKU children tend to show slightly but significantly less verbal ability than otherwise identical control groups. The critical issue, however, is that carriers, like the normal population, *never have elevated blood levels of phenylalanine,* the amino acid that causes profound retardation in their children. Scientists thus assume that the PKU gene must also directly affect certain parts of the brain connected with verbal IQ.

Sufferers of PKU are identified by a test of their blood or urine. Until recently, because carriers had no distinguishing marker, they could only be identified if they had produced a PKU child. On this basis, estimates have placed the number of carriers in the United States alone at about two and a half million.

But a reliable blood test that identifies carriers has recently been developed. A single midday, premeal sample of blood is enough to identify those who have one PKU gene. Now, according to Charles Scriver at the McGill University-Montreal Children's Hospital Research Institute in Montreal, Canada, "It is possible to predict and prevent the genetic and social effects of PKU and its variants."

IQ and ABO Blood Groups. Tests of the blood groups of the inhabitants of seven English villages in 1971 revealed that people who had blood type A_2 (a variant of blood type A) had a slight but significant advantage in intelligence over those with other blood types. The link between A_2 and IQ was strong enough so that, *in that particular homogeneous population,* it could be used to predict somewhat higher scores on IQ tests. While other populations might not show the same correlation with the A_2 blood type, the finding was important; it implies that blood factors can be linked to the groups of genes that influence IQ. Just as certain markers that reveal predisposition to cancer appear only within families (demonstrating a purely familial link between the marker and the gene or group of genes that causes the susceptibility) many genetic factors may be connected to IQ within homogeneous populations. Ultimately, they may surface as markers; and we may be able to predict slight differences in IQ *within populations* based on the links between intelligence and specific genetic traits.

Markers like PKU, Fragile X, and blood type A_2 demonstrate the connections that exist between genes and the elements of normal intelligence that are measured by IQ tests. They point out the fact that some patterns of intelligence are predictable and solidly linked to heredity. As more markers appear, the accuracy of our predictions will increase.

But what of the crucial question, the possibility that, based on what we now know, we can compare the intellectual capacities of different groups and consign some to inferior status? That, of course, is the contention of some. They use the evidence gathered by scientists like Arthur Jensen, who recently published a massive document titled *Bias in Mental Testing* to support his view that the IQ tests on which blacks score an average of fifteen points lower than whites are, in fact, valid.

Jensen's book has not sparked the expected explosive debate partly because the scientific community has turned away from the idea that IQ can be a legitimate way of predicting anything. We know now that IQ scores mean very little in determining who will earn more money and who will earn less. We have discovered that environmental factors can bounce the results of the test up and down like a yo-yo. And we are beginning to make inroads into the workings of the brain itself, isolating the areas in which thought and reasoning, memory and calculation reside, learning the biochemical roots of the mind. In the context of this new knowledge, IQ tests diminish in value.

Volumes have been written both supporting and attacking Jensen's original conclusions. Perhaps the most telling statement of all was released by the Genetics Society of America, which, in 1976, resolved:

> It is particularly important to note that a genetic component for IQ score differences *within* a racial group does not necessarily imply the existence of a significant genetic component in IQ differences *between* racial groups; an average difference can be generated solely by differences in their environments. . . . In our views, there is no convincing evidence as to whether there is or is not an appreciable difference in intelligence between races. . . . We feel that geneticists can and must also speak out against the misuse of genetics for political purposes, and the drawing of social conclusions from inadequate data.

In the final analysis, whether or not Jensen's contentions are true is irrelevant. It is probable that differences in the genetic bases for intelligence do exist among racial, sexual, and other groups, just as differences exist over the entire range of physical characteristics. It is even possible that the different scores of the various races point to associated genetic variation. Then again, they might not; we simply have no way of knowing what those differences mean. Our knowledge is, in fact, still so limited that *any* theory about the relative intellectual capacities among races cannot be proved or disproved at this time. Those who claim they have solid evidence to the contrary are taking an enormous, insupportable step from IQ test results to racial theories; theirs is a leap of faith, not of science.

Genes and Personality

Intelligence is the most noticeable, most measured, most contested aspect of normal behavior and personality. Other personality traits have been neglected in the continuing battle over intelligence, the trait for which the stakes are highest. Nevertheless, some scientists have been working with everything from sociability (the tendency to be introverted or extroverted) to levels of energy, from emotions to the drive for achievement. Not surprisingly, the evidence they have been compiling is consistent with findings on every other part of the composition that is man: There seems to be a clear genetic influence in each basic element of personality.

Perhaps the most unusual discoveries have come in the area where many scientists expected to find the slightest impact of inheritance. The degree to which we are introverted or extroverted would seem, at first glance, to be one of the traits *least* likely to be influenced by genetics, almost certainly guided by culture and upbringing. But psychological tests and twin and adoption studies all bear each other out: one's ability to get along with others has a definite genetic component.

Nobody really knows why this should be so. One theory has it that the extremes of sociability were bred out of humans millennia ago; being too gregarious when you should have been out looking for food would have been frowned on, while being too much of a loner when men needed to hunt together in packs would have been unproductive. Those who were better hunters or gatherers were more highly prized for reproduction; and, gradually, sociability—perhaps an insignificant trait originally—strengthened its connections with the genes.

Support for this thesis has now come from research performed by Joseph Horn of the University of Texas and Robert

Plomin of the University of Colorado. The two scientists began by undertaking a project that seemed only remotely related to the issue of the genes' impact on personality: they were looking for personality traits that could predispose people to heart disease.

Horn and Plomin studied the results of personality tests given to 200 sets of twins drawn from the registry of 15,900 kept by the National Research Council. The twins, all veterans of the military, were middle-aged men. Different sets had been evaluated for the role of smoking in heart and lung disease, the effects of air pollution, the genetics of psoriasis and multiple sclerosis, and the causes of early death. This study of behavior and heart disease was supposed to be just another in a long list of research projects in which they were involved.

Horn and Plomin began by comparing the tests of fraternal and identical twins; and they found about 50 personality factors (aggressiveness, introversion, extroversion, and others) for which identical twins were more alike than fraternal twins, implying that those factors had some genetic component. Their heart study forgotten, Horn and Plomin began to examine each factor, trying to determine what elements went into it, struggling to find a common thread that might hold them all together.

When the dust had cleared, one trait stood out. It was characterized by the ability to talk to strangers.

Plomin calls the trait "gregariousness." Horn labels it "conversational poise." No matter what it is called, it has a clear parallel at one point in an infant's neurologic development—its response when it begins to acknowledge the presence of strangers in its environment. Nobody knows why different infants react differently, but it may depend on whether they experience the situation as pleasant or painful. And the perception of that sensation may be under genetic control.

Sociability is not the only personality trait that seems to have a genetic component. Other researchers are looking for similar results in tests for emotionality, energy level, impulsive (and its opposite, predictable) behavior, and general levels of activity and drive. So far the results are inconclusive.

No genetic pathways have been worked out for any of these traits; whatever they are, they are probably complex and interactive, depending on groups of genes rather than any single genetic factor. Markers that may be found will probably predict tendencies and minor shifts in personality, rather than any major predispositions to specific types of behavior.

Nevertheless, at least one physical characteristic has already been linked for predisposition to certain behaviors. Its existence may indicate that the physical attributes that make an individual better or worse at certain activities will tell us more about someone's future than minor shades of difference in personality.

About a decade ago, Morgan Worthy of Georgia State University noticed that certain athletes performed differently in different sports. The differences seemed to be tied directly to the type of activity each sport required, and specifically to whether the athlete had to react at top speed to sudden changes, or whether he or she could develop a personal rhythm in the sport. Through observation, Worthy found a correlation between eye color and activity; athletes with darker eyes seemed to respond better in situations requiring split-second timing, such as hitting a baseball or boxing; athletes with lighter eyes seemed to do better in sports and positions that required responses to relatively stable situations, such as golfers and baseball pitchers. The trait was not tied to race. The same distinction developed between dark and light-eyed Caucasians as between blacks and whites.

At first the correlation seemed absurd. But Worthy took the theory one step further, conducting studies of animals with light and dark eyes. He found that the same correlation existed: "reactive" behavior in animals seemed to go hand in hand with dark eyes; "self-paced" behavior was linked to lighter colors.

Surveys of athletes tended to support these findings. Worthy and his colleagues found that blue-eyed bowlers actually win more money on the professional tour than those with brown eyes; and dark-eyed people tend to excel at what is often termed the most difficult skill in sports: batting a thrown baseball.

Soon other researchers began to look for more definitive experimental evidence to support or contradict Worthy's assumptions. Among them, Peter Post of New York Hospital designed an experiment to test the reflexes of light- and dark-eyed subjects under laboratory conditions.

Post designed a contraption that would drop a ruler at the touch of an electronic switch. The volunteers in the experiment had to try to catch it as it fell. The test took into account such variables as sex, the time of day the test was given, and whether the subject was left or righthanded. It tested people's reflexes under three separate sets of conditions: with the ruler dropping alone; with the drop accompanied by the ringing of a door buzzer; and with the ruler dropping as a light flashed.

The results gave powerful support to the theory. Major differences were found when the drop was accompanied by a flash of light. And both males and females with the darkest iris color—medium brown—were faster than those with lighter colored eyes.

The reason for the difference is still unknown. One theory is that the increased pigmentation in darker eyes enhances the speed of neural messages from the eye to the brain.

The genetic basis for eye color, then, might influence behavior indirectly, through its impact on the ability to react. Clearly, people with faster reactions confront an environment that is slightly different from what it is for those with slower reactions, just as someone who is color-blind perceives a different environment from someone with full-color vision. How these tiny alterations might affect behavior is as yet unknown; the complex nature of behavior may mean that countless variations are absorbed into the whole personality without leaving much of a trace. But it might also mean that the more important variations can cause significant shifts in our behavior and our lifestyles. Certainly, eye color is not the deciding factor in an athlete's decision about what he or she should be doing; there are capable brown-eyed pitchers and blue-eyed batters all over professional baseball. But if it tips the balance, if it makes a difference in the quality of an athlete's performance, it may be one of the elements that helps determine profession, attitude, and the extent of one's skills.

One major danger with these preliminary findings is the possibility that they might be stretched too far. Defining an athlete as "self-paced" simply because he has blue eyes is forgetting the hundreds of other elements that may be more important in his choice of profession. Looking at statistical differences among IQs and extending them to predict differences among races ignores the impact of cultural and environmental factors, as well as the possibility that different races may actually inherit different aspects of intelligence. We have come a long way since the time when scientific findings were accepted as gospel and related directly to goings-on in society. But there are still times when scientific evidence may look incontrovertible, may be used, and may turn out to be false.

The Extra Y: A Criminal Chromosome?

On July 13, 1966, a tall loner with "Born To Raise Hell" tattooed on his arm brutally murdered eight Chicago nurses in their residence. Soon after, Richard Speck was captured, tried, and imprisoned. Journalists covering the case had a field day. Several self-appointed genetic experts speculated loosely about his genes. They noted his height and his criminal record and suggested that he might belong to a class of criminals that had been discovered just the year before: a group of unusually tall men who carried an extra Y chromosome in addition to the normal male set of XY sex chromosomes.

High rates of XYY carriers in mental-penal institutions had been discovered in Edinburgh, Scotland. Subsequent studies confirmed the initial findings: The incidence of XYY chromosome carriers in the general population is about one in a thousand; but in institutions, it was found to be about two in a hundred. That difference was considered significant. Scientists and journalists alike theorized that the abnormality was a marker for criminal behavior; several lawyers tried to defend their XYY clients by pleading insanity based solely on the presence of the extra Y.

Richard Speck turned out to have the normal complement of chromosomes, and the furor surrounding the journalists' suspicions soon died out. But once questions about the criminal tendencies of those with XYY chromosomes had been raised, they had to be answered. In 1968, two Boston-based scientists, Stanley Walzer, a psychiatrist, and Park Gerald, a geneticist, decided to survey all male offspring born in Boston's Children's Hospital Medical Center, typing them for their sex chromosomes and following those who turned out to have more than the usual complement.

The project began quietly enough. But by 1973 it had aroused a storm of controversy around the Boston area. In an ideal world, the project might have been a fine bit of scientific investigation; but in a world that had managed to tag carriers of XYY as holders of the legacy of Richard Speck, it seemed dangerous.

Those who opposed the research raised serious questions both about its value and about its impact on its young subjects:

- Had the devastating publicity given the XYY type crippled the Walzer study even before it began because of the "criminal" label that would be applied to anyone with the chromosomal abnormality?
- Did the scientists' methodology, which involved informing the parents of the abnormality, render the study useless by changing parental behavior toward the offspring and affecting the critical issue of the environment?
- Was the study endangering children who might otherwise grow up to be normal, leading to a kind of self-fulfilling prophecy?

There were no good answers to these questions. Clearly, at least some people with the XYY type grew up to be acceptable members of society. And there was the chance that the children might be adversely affected. Nevertheless, when the question went before a board composed of members of the Harvard University faculty, Walzer managed to convince them that all necessary precautions had been taken; the study was approved.

Soon after, the crisis deepened. The opponents of Walzer's research claimed that some members of the board had been pressured into voting for the study. Walzer himself claimed that his family had been threatened in anonymous phone calls. Ultimately, citing the destructive effects of this kind of pressure, he suspended his study.

The demise of Walzer's study did not mean that all investigations into the effects of the XYY genotype were halted, however. Society still had an overriding interest in the truth. If XYYs were, in fact, prone to committing aggressive crimes, and did so out of all proportion to their numbers in the general population, society had a stake in knowing it and trying to respond to it, no matter what ethical and moral issues might be dredged up. And if the statistics were wrong, XYYs around the world were on the verge of being unjustly harassed.

In 1976, a study performed in Denmark cleared up most of the controversy. The investigators examined the chromosomes of thousands of tall Danes (since those with the abnormality are, on the average, six inches taller than the general population) so that they could study a group of XYYs chosen from among ordinary citizens, rather than from among inmates of an institution. Then they took case histories to find out whether those in this sample had been convicted of crimes, evaluated the types of crimes, and examined their findings in the light of three possible explanations: that XYYs were, in fact, more aggressive and antisocial than the rest of the population; that the intellectual impairment known to be brought on by the extra Y chromosome made XYYs easier to catch; and that their added height made aggression easier and caused others to perceive them as more dangerous.

The findings went a long way to clearing the air. The Danish group found that about 42 percent of the XYYs they uncovered (five out of twelve) had criminal records, much more than the 9 percent who turned up among the controls. But except for one man who badly beat his wife while he was drunk, the crimes were neither aggressive nor violent. And some of the convictions were directly tied to crimes that someone with normal intelligence and impulses would never have been caught doing. One man took to burglarizing houses when the owners were around; another was arrested for calling in a false alarm for a traffic accident.

The study concluded that the extra Y chromosome does offer higher risk of antisocial behavior. But the risk is much lower than was originally thought. And there is no evidence that the added sex chromosome is tied to aggression. In fact, XYYs in prison are now believed to be *less* aggressive than their XY counterparts.

Animal Genes

As the controversies over both intelligence and the significance of the XYY genotype make clear, studies linking genes and human behavior are particularly difficult to undertake. The ethical constraints against experimenting with human beings under laboratory conditions is one reason; the fantastic range and richness of variation within our genetic heritage is another. To compensate, researchers often count on animal studies to provide analogies for the human condition. More often than not, they unearth amazing genetic evidence.

Investigators have managed to breed mice that show a preference for alcohol over water, or, for that matter, for water over alcohol. There are rats that are genetically predisposed to work harder to get alcohol, even when water is freely available. Dogs can inherit a propensity for cowardice. Guinea pigs have a genetic predisposition against elaborate sexual foreplay. Numerous animals show inherited sensitivities to harsh sounds, cold, gravity, salt, and certain odors.

Sometimes it seems as if almost anything that a scientist wishes to discover will appear in the genes of one animal or another. Geneticists have actually bred a rat that carries the genetic tendency to kill mice, for instance. The "muricidal rat," as it is called, acts normally until a mouse appears in its cage. Instantly, it turns ferocious; it pounces and tears the intruder apart. Mice may have similar "killer genes." They are normally good mothers; but they can be bred to rip huge chunks of flesh from their offspring, not eating them, as guinea pigs often do, but dismantling them, piece by piece.

The problem, however, is not in the scientific findings themselves, but in how we interpret them as we decide what is relevant to humans and what is not. In each case, we have to determine whether a study is applicable to man; whether humans, too, might carry a gene or group of genes that predisposes us to vicious murder or infanticide. Scientific studies of the genetic basis for animal behavior are important; but they must be looked at in the light of human genetic and environmental variation. Their real value lies in their ability to direct our attention to what we should be looking for in people and not necessarily in the discovery of a specific trait that might happen to surface in a species of animal.

The differences between gene expression in laboratory animals and human beings are profound. Animals in test situations have had both their environment and their genetic composition carefully controlled for generations. When a scientist searches for a single specific characteristic among them, the research is like taking a blank wall and painting a swath of red through its center; against the neutral background the scientist has carefully prepared, the red stands out clearly. But human genes and environments offer a wall swirled with color. That same bold shade of red may disappear among the powerful patterns already there. And scientists have neither the time nor the moral right to systematically strip that wall down, color by color, to find out what is in control.

Animal studies, then, function as metaphors for the human condition. They must be accepted for what they are; experiments that isolate genes in an environmental vacuum. By themselves, they cannot tell us whether, or to what extent, similar traits exist in humans.

Nevertheless, the weight of evidence in both animal and human studies points to the power of genes in behavior. Does it necessarily follow that we will one day be able to predict personality, intelligence, or potential areas of neurosis all on the evidence provided by the genes? With the constant shifting of factors, the interplay between genes and environment, probably not. But genetic markers will enable us to predict *tendencies*. As the biochemical, genetic, and environmental foundations of behavior are unearthed, we will certainly be able to predict an individual's predispositions—the likely occurrence of some general parameters of behavior—given a specific set of circumstances. The fundamental question that remains is: Is that really what we want to do?

11

THE PERILS OF PROPHECY

A new scientific truth does not triumph
by convincing its opponents and making them see the
light, but rather because its opponents
eventually die, and a new generation grows up that
is familiar with it.

MAX PLANCK

Each new power won *by* man is a power
over man as well.

C. S. LEWIS

The tiny farming village of Orchemenos in Greece must have
suffered heavily from malaria throughout its history. That is
probably why, by the late 1960s, about a quarter of its entire
population still carried the sickle-cell trait, and one in one
hundred of its babies was born with sickle-cell anemia. In
Orchemenos, the sickle-cell gene had offered a practical trade-
off: It killed about 1 percent of the population but allowed a
substantially larger group to resist a deadly disease.

When scientists learned of the high incidence of sickle-cell
anemia in Orchemenos, they recognized the dual opportunity
it offered them. They could help the villagers avoid marriages
between carriers of the sickle-cell gene by instituting a carefully regulated program of genetic screening and counseling;
at the same time, they could explore the effects of such a
screening program on a homogeneous, controlled population.

The customs of Orchemenos made the experiment even more plausible. Over half the marriages in the village were arranged in a traditional fashion, between patriarchs of families. And the health of the two potential partners was a critical criterion in each arrangement. Identifying carriers of the sickle-cell trait would add some new and important information to the discussions. The scientists hoped that they could convince the villagers to take their genotypes into account in their arrangements. If carriers of the sickle-cell trait could be persuaded to marry people without the gene, the village could be wiped clean of the sickle-cell disease.

Each villager underwent screening. The entire village was counseled as to the meaning of the results. Then the scientists left the village alone. Seven years later, they returned to find out what had happened.

What they discovered was more or less a complete surprise. As they had hoped, the villagers were including the presence or absence of the sickle-cell trait in their discussions. But despite the scientists' attempts at education, those who had been found to carry the trait were being squeezed out of the mainstream; the rest of the population now considered them "inferior"—even though there was little chance that they would ever suffer from the effects of the gene they carried. Those without the sickle-cell trait now tended to marry among themselves. So did the carriers. The carriers' isolation was so pronounced that the number of marriages between carriers was just as high as it had been before the screening program had been instituted. The new knowledge had not led to enlightenment; it had not reduced the incidence of sickle-cell anemia in Orchemenos; it had not spawned a community-wide effort to eradicate a tragic and fatal disease. Instead, it had become just another way for the villagers to classify one another. It had created a new and stigmatized social class.

Orchemenos is a tiny rural community in which everybody knows everybody else's business, a backwater awash in tradition. When the soothing mist of ignorance that had surrounded its view of sickle-cell anemia suddenly disappeared, the community had nothing to fall back on for support. The villagers could sit in their small counseling sessions and nod as if they understood what they were being told. They could even integrate the information into their way of life. But without a firm foundation of prior understanding, they could not fully accept the meaning of what they were being told. The concept of inherited *differences* came to mean inherited *deficiencies*. And so they instinctively created a kind of mythology around the meaning of the sickle-cell trait, a mythology that specifically contradicted all that the scientists had told them. The careful program of screening and counseling had somehow failed. It had done little to change the balance of health and illness in Orchemenos. The village proved unable to help itself. The availability of a genetic marker left the inhabitants no better off than they had been before.

Genetic prophecy is a powerful tool for the future. But it is just that, a tool. It can expand the scope of our knowledge; give us clues as to what lies ahead; help us map strategies that offer better odds for healthier living. But it is not simply a new machine or an innovative idea, a better, faster, more efficient way of doing the kinds of things we have always done. Genetic prophecy literally has the power to change the way we live. It will introduce new priorities into some of the most fundamental decisions we have to make—about the profession we choose, where we live, whom we marry. It will, in short, cut to the very core of our lives.

The experience of the village of Orchemenos illustrates what can happen when that kind of scientific power is let loose on an unprepared community. It is just another in a long list of examples of what can happen when we gain the ability to do something and begin to use it before we understand all the repercussions. We have learned, for instance, that nuclear power can provide relatively inexpensive and plentiful energy, but that it may place nearby communities in serious jeopardy. We have discovered that the indiscriminate use of miracle drugs has actually triggered the evolution of drug-resistant super germs. We have found that producing the space-age materials we count on so heavily places a strain on natural resources and puts those in the relevant industries at increased risk for disease. In each case, we have learned that important advances are not entirely beneficial. They involve trade-offs, an exchange of enhanced risks for their benefits. By now, we generally accept a basic rule that seems to accompany almost every application of scientific solutions to societal problems: the greater the opportunity for benefits, the larger the potential for abuse.

Now we are facing an extraordinary revolution in our way

of providing health care. We are on the verge of a truer understanding of the nature of disease. We are witnessing a massive shift away from an emphasis on curing diseases after they strike and toward preventing them before they occur. Genetic prophecy is in the vanguard of this movement; and it contains its share of risks.

The perils of prophecy do not come from the science itself; the act of taking a sample of blood or urine and analyzing it for its components is quite safe as long as we understand exactly what that analysis means. The real risks come from applying the information, from the ways that genetic prophecy can be misinterpreted and misused.

If we are not careful—if we look only at what prophecy can do *for* us, and not at what it might do *to* us—we may lose track of the roles genetics has played in the past. If we ignore the struggles that have occurred in Nazi Germany, the eugenics movement, the racial history of the United States, we will be guilty of exploring genetic prophecy in a vacuum, of failing to take into account the social, ethical, political, and moral dilemmas that influence the manipulation of power every time science meets society. The effects of prophecy are sure to be widespread. Unless we examine them before they can take hold, we are virtually guaranteeing that abuses will occur.

To Screen or Not To Screen?

The concept of medical screening is not new. Centuries ago, private physicians tested their wealthier patients for diabetes by tasting their urine to evaluate its sugar content. Quarantines were placed on entire towns in which epidemics were spreading. Today, nearly everyone who applies for a marriage license is screened for syphilis, and women routinely undergo yearly PAP smears for signs of cervical cancer. Nevertheless, screening has, for the most part, been limited to detecting infectious diseases in their early stages and, during the past decade, to some basic prenatal testing.

Genetic prophecy will break through these limitations. Instead of identifying patients, it will locate *potential* patients; instead of screening for disease, it will search for *susceptibility* to disease. Prophecy in full flower means that practically anyone will be able to learn of the diseases he is predisposed to contracting. The various screening programs will affect everyone, even children not yet conceived (who can be characterized through their parents' genes). They will make it possible for people to order their lives around the important questions of health and safety that apply specifically to them, if they wish. They may, in the end, change the very character of society.

Mass screening is the key to the future of genetic prophecy. It is based on two technological breakthroughs: the development of computerized testing procedures that will enable us to take a sample of blood or urine and analyze it for literally hundreds of different components; and our ability to identify genes directly by using the new techniques of biotechnology. Through mass screening, physicians will be able to test entire populations for a few dollars per person. Predictions that are now offered on the basis of relatively few experimental returns will be based on results found among literally millions of people from different racial, geographical, and ethnic groups. The art of prediction will be refined to the point where an individual's identification with various groups, along with the genes that he carries, will pinpoint the risks he faces from specific environmental conditions.

The potential of mass screening was obvious from the very beginning. Within months after the first primitive test for diagnosing phenylketonuria (PKU—one cause of mental retardation) was developed in 1963, the state of Massachusetts had passed a law that required that all newborn babies be tested for the disease. Within a decade, 42 other states had drawn up similar legislation. Together, they created a legal patchwork, an ill-advised outpouring of solutions that ignored the potential problems of screening programs as they pushed society toward what some considered genetic salvation. Some states provided funds for treatment programs; others did not. Some offered counseling; others simply told the parents of the test results. Few seemed to recognize that screening for genetic diseases presented different problems from screening for infectious diseases. And some legislators did not even realize that PKU was a genetic disease.

After the first wave of legislation had passed, critics of the programs began to take stock. The test that had been developed turned out to be less accurate than people had believed and was giving 19 false positive results for every real case of

PKU it uncovered. As a result, some infants were being placed on highly restricted diets that actually caused them harm. The disease is also restricted mainly to people of European origin. The District of Columbia spent three years and over $100,000 without turning up a single case of PKU in its largely black population; in 1971 its screening program was suspended. Finally, the treatment that many assumed could cure those with the disease turned out to be imperfect; although children with PKU who were treated were vastly improved over those who were not, their IQ scores remained slightly lower than those of the general population. They could not be completely protected.

Nevertheless, even though most observers criticized the lack of planning and foresight that had gone into the various PKU screening programs, the general consensus supported the principle of screening. PKU was a clear-cut disease. It could be treated if it were diagnosed early enough. Gradually, the screening process was improved; some of the laws were modified; and every state that had begun a screening program in the wild optimism of the 1960s continued it in some form.

Unfortunately, the lessons of the PKU experience had not sunk in before the second round of screening began in 1971. And this time, the benefits were not nearly as obvious. The disease in question was sickle-cell anemia. The mass screening programs were designed to identify not only those who had the disease—a problem for which there was (and still is) no cure—but also those who were merely carriers of the sickle-cell trait. In the minds of both the public and various legislatures, the group at risk was the entire black population, a group that had had its fill of being isolated during American history.

The movement toward sickle-cell screening was actually triggered by black militants who, in the late 1960s, contended that the neglect of the sickle-cell problem by the federal government was a direct result of discrimination and racism. Their voices found the sensitive ears of politicians tuning up for the general election of 1972. President Richard Nixon responded by requesting massive funding for sickle-cell research and focused national attention on the problem.

Many states took Nixon at his word. Again, Massachusetts became the first state (of thirteen) to draft a mandatory screening law. It also declared the sickle-cell *trait* a disease, ignoring years of medical opinion that had found sickle-cell carriers to be at risk only under specific and relatively uncommon environmental conditions. The City Council of the District of Columbia decided that sickle-cell anemia was a communicable disease, a characterization that is technically correct but that enabled others to confuse the passing of genes from parent to child with the passing of infection from person to person. In all, about 30 states began the process of legislating screening programs for sickle-cell. And those who made their programs mandatory required blacks—and only blacks—to undergo screening either before they entered the public school system or when they applied for marriage licenses.

The differences between screening for PKU and for sickle-cell disease were enormous. Basically, people who discovered that their children were threatened with severe brain damage by PKU had options; the disease could be spotted before it caused problems and could be treated. People who were identified as having sickle-cell had little they could do with the information. There was then no scientific test that distinguished between those who were merely carriers and those who actually had the disease. There was no prenatal test that could tell whether unborn children had the disease. There was also no cure. The only value of the screening process was to alert healthy people—those who carried only one copy of the sickle-cell gene—to the risks their children would face if they mated with another carrier. And the only option they had available to them was to choose marital partners on the basis of a laboratory test.

Perhaps the sickle-cell screening program would have been beneficial to the community had the states placed emphasis on counseling, on taking care to teach blacks exactly what the test results implied. But that was an aspect of screening that most state laws ignored.

Curiously, at almost the same time, another identifiable group was undergoing screening, and almost without controversy. Many Ashkenazy Jews carry a particularly deadly trait: the gene for Tay-Sachs. Tay-Sachs is a recessive disease that leaves carriers unharmed but destroys the nervous system of those who inherit two such genes, inevitably killing them within the first few years of life. One in thirty Ashkenazy Jews carries the gene; the odds of two of them marrying are one in 900; and the odds of any one of their children getting Tay-Sachs are one in four. The chance, then, for any one member of the community coming down with Tay-Sachs is about one in 3,600.

Fortunately, Tay-Sachs carriers can be identified; a prenatal test exists that can tell if their children have the disease. In the early 1970s, many Jewish communities set up voluntary screening programs so that Tay-Sachs fetuses could be found and aborted before birth. The program worked beautifully for several reasons: the test is accurate and reliable; the disease is terrible enough to erase most people's doubts about the morality of abortion; the programs stressed community involvement; and most Jews were well educated in the nature of Tay-Sachs. The screening program took the fear out of giving birth for many couples. By preventing the births of several hundred Tay-Sachs babies, it probably also *encouraged* the births of hundreds of normal children, children who might not have been conceived either because a couple had already had a Tay-Sachs baby or because they feared that possibility.

The differences between sickle-cell screening and screening for PKU and Tay-Sachs were lost on very few. Critics who had toned down their disapproval of the PKU program because of its obvious benefits went after sickle-cell legislation with a vengeance. While many acknowledged that the programs meant well, few were willing to let them stand without meaningful revisions:

- Most laws that had been enacted did not even begin to protect the confidentiality that is the right of all patients. In some instances, the records of those who had undergone screening were available to practically anyone who wanted to see them.

- The failure of most laws to provide comprehensive genetic counseling meant that those who discovered they were carriers had no way of finding out what that entailed. As a result, many began to shoulder a burden of guilt, a sense of genetic inferiority, a belief that the presence of the single sickle-cell gene somehow made them defective.

- The information offered by the screening test was clearly not helping those identified as carriers. On the contrary, they were being harassed—subjected to higher insurance rates and discriminatory hiring practices, all because of their ties to a disease that they simply did not have.

It was then that mass genetic screening approached the first crossroad of its short, turbulent career. The criticisms had to be answered. Those who had the power to do so—professional advisory groups and legislative bodies, for the most part—were confronted with two clear choices: Either they could modify the screening programs to protect those being screened from the potential for abuse; or they could allow things to continue largely as they had, instituting programs for other diseases without regard for the risks that they might offer.

The Abuse of Power

Imagine the possibilities: An executive on the rise is invited to lunch with his company's chairman. There he is given the bad news: He is considered a terrific worker, a bright and imaginative leader. But his medical records show him to be at risk for heart disease. Because the company cannot take the chance that he will die young, he will not be considered for promotion. The chairman is sorry, but other people with excellent qualifications are available. And they don't carry the baggage of high risk.

The Congressman from an important industrial state has decided to run for higher office. He is forceful and articulate, well liked in his district, and the senior Senator is about to retire. The day before he declares his intentions, he receives an anonymous telephone call. His medical files have been rifled. The opposition has discovered that he carries a gene that predisposes him to manic depression. If he announces that he will run for the Senate, that information will become public knowledge. The Congressman decides not to run.

A couple decides to have a baby. They discuss it with their family physician. He tells them that their particular genetic patterns make it likely that they will have a child who will indulge in a life of crime, without a chance of finishing school, with a high probability of mental or physical defects that will cause it to die prematurely. The odds are one in four, high enough so that they are denied the opportunity, by law, to take the chance. He is sorry but there is nothing he can do.

Knowledge is power. In genetic prophecy, it offers choices where none existed before. But power can be abused as easily as it is used. When a single sample of blood or urine divulges information not only about someone's physical susceptibilities but about his or her propensity for psychological problems as well, it becomes a choice piece of carrion for the vultures and hyenas of the world. It is, in fact, a scientific advance that offers access to the explosive medical secrets of whoever is screened. One paradox looms large: By seeming to give each person more control over his or her destiny, genetic prophecy creates the possibility that some other group will use the knowledge to take that control away.

Part of the power of prophecy stems from its presentation of difficult value judgments. When an individual makes his own choices—in deciding, for instance, to continue smoking in the face of overwhelming odds that he will contract lung cancer—these judgments may be acceptable. But when the choices are made by others—by an employer, an insurance company, another member of his family, the government— serious problems arise. Knowing about someone's predisposition to a mental disorder is only a short step away from using that knowledge prematurely or dangerously. With widespread genetic screening, we may find ourselves faced with a Big Brother who works to direct each individual's future by tapping in to his genetic fate whenever he sees fit.

The possibility that this could happen is not as remote as it might at first seem. Medicine has long been used as a rationale for political ends. In the Soviet Union, political deviants are routinely turned over to physicians for confinement and treatment for "mental" disorders. And in the United States in 1964, nearly a thousand physicians signed a statement pronouncing the Republican candidate for President, Senator Barry Goldwater, "unfit" for the job he was seeking in an assessment that had nothing to do with the Senator's state of health. Genetic prophecy multiplies the potential for these kinds of abuses.

Fortunately, obvious abuses often have obvious solutions. In December, 1971, a group of geneticists and ethicists led by Marc Lappé of the Institute of Society, Ethics and the Life Sciences in Hastings-on-Hudson, New York, met to discuss possible solutions to the problems raised by both the PKU and the sickle-cell screening programs. They drafted a set of guidelines that would effectively curtail most institutional abuses that might arise.

The group's guidelines dealt principally with the issue of personal rights and freedom. They called for voluntary participation in screening programs; the informed consent of the person being screened; free access by the participant to the information gathered; extensive counseling programs that would offer insight into the meaning of the test results—and that would avoid recommending specific courses of action; counseling and informed consent for any treatment programs that the results of the screening might call for; absolute secrecy surrounding the information gathered, similar to the doctor-patient confidentiality that is the core of most medical relationships; and education as one of the primary goals of any screening program.

The guidelines of Lappé's Genetic Research Group were published in the *New England Journal of Medicine*. They were quickly adopted into the National Sickle Cell Anemia Control Act of 1972 and the National Genetic Disease Act in 1976—laws that brought the federal government into the fray. The National Sickle Cell Anemia Control Act authorized $80 million for the development of programs that included screening, counseling, and public education; the law required that the tests be voluntary and confidential, with substantial community involvement. The National Genetic Disease Act widened the law to include other genetic problems. It also made federal aid available only to those states that enacted purely voluntary programs.

Maryland and California incorporated the Genetic Research Group's guidelines into their own laws. Maryland went even further, setting up a sixteen-member commission composed mostly of laymen to regulate the state's own screening programs. Other states were less willing to comply; the Maryland safeguards raise the costs of screening programs substantially.

The new laws have resulted in an enormous increase in federal and state support of mass screening programs, especially for prenatal and newborn babies. At least a fifth of all states can now screen infants for a variety of genetic diseases in addition to PKU. At least 40 percent provide some genetic counseling (although Virginia's contribution is limited to covering the expenses of one nurse, Arkansas manages to get by with paying half the salary of a medical geneticist, and Wisconsin budgets enough for one genetic social worker). Most states make provisions to preserve the confidentiality of genetic information.

Safeguards like these will help limit the blatant abuses of genetic prophecy by making it difficult for governments, employers, insurance concerns, and other interested parties to break the seal of secrecy surrounding any one person's genetic heritage. Still, other abuses may occur. Insurance companies and employers, for instance, could extend their mandatory physical examinations to include genetic screening; for the *risk* of illness is the whole point of these testing programs. Insurance rates are already almost entirely governed by an assessment of risk; the lower rates for safe drivers and nonsmokers are based on statistical evidence that these people are better insurance risks (although monitoring the truth of someone's claim that he or she does not smoke may be next to impossible). Genetic prophecy offers perhaps the best way of assessing the chances of someone remaining healthy in the future.

Already, the prospect of more accurate prophecy is making some people take notice. The Standard Asbestos Manufacturing and Insulating Company, for example, is suing cigarette manufacturers for the high cost of *its* insurance, pointing out that a worker who smokes is one hundred times as likely to contract asbestosis than one who doesn't and claiming that it is not the company's responsibility to assume the cost of the additional risk.

On the other side of the coin, an editorial in the *Medical World News* of May 18, 1979, supported a bill introduced by Senator John Danforth of Missouri that proposed that catastrophic health insurance be paid for by raising the federal tax on cigarettes. Perhaps programs for alcohol abuse should be supported by an added tax on liquor and car-safety research by a higher penalty for speeders. Today, the editorial noted, nobody pays fully for his or her medical care; the burden is shared by government, employer, and other taxpayers: "Illness nowadays is a social event, a social expense. This means people who try to take care of themselves are subsidizing the big medical bills run up by reckless drivers, heavy smokers, big eaters, and other immoderates. It's about time to bring the cost of such foolhardy habits home to roost." In other words, if someone knows that he is at risk and persists in taking chances, why should society be forced to foot the bills?

As safeguards against some abuses are instituted and discussions about others continue, the likelihood that mass genetic screening will suffer blatant misuse becomes more remote. Yet many questions remain unanswered: some tests falsely identify healthy people as potentially ill, and those at risk as normal, far too often; others provide information that we can do nothing about; still other tests exist for diseases that are so rare that screening huge populations for the one person in forty thousand with the syndrome is not economically feasible. Nevertheless, the *principle* of mass screening is well established. Its potential benefits cannot be ignored. Its use is practically inevitable. And that is the framework within which other problems should be discussed.

Stereotypes and Stigmas

The confidentiality that is so vital to a safe screening program protects the identities of those being tested. But it does not preclude researchers from using the general results of the tests to reach valuable conclusions about entire populations. If those in charge of the PKU screening program in the District of Columbia had agreed to shred the results of each test after informing the individual involved, they might not have discovered how useless their testing was, given the racial composition of the District's population. They might have continued to pour money into a program that had virtually no chance of offering a return on their investment.

The more we learn about genetically distinct groups, the better we are able to characterize them, to recognize which genetic traits they tend to carry. Dividing populations on the basis of their genetic heritages can be helpful. We can begin to tailor screening programs to meet a particular group's unique requirements, enhance the programs' effectiveness and reduce their costs by concentrating them where that can do the most good—we can even suggest changes in the way some groups live, so that they can take advantage of their genetic strengths. But the danger of misinterpretation still exists.

Because of it, many people are against the very idea of genetic research, contending that as we discover the power of the genes, we are providing fuel for morally insupportable and politically dangerous points of view. Other critics suggest that we preserve our ignorance of the relationship between genes and our fates so that we can sustain our faith in the concept of free will. The argument has its attractions. But it is reminiscent of a debate that Abraham Lincoln once had with his generals. When he could not convince them of the facts he presented, he asked them: "How many legs does a sheep have, if you count the tail as a leg?" Five, they answered. "Sorry," said Lincoln. "Counting a tail as a leg does not make it one."

Biologically, we are not created equal. There is nothing fair about heredity. The genes we receive are not parceled out by political doctrine or law. One population may be superior or inferior to another for just about any genetic trait that we manage to measure, even though, on balance, they may all turn out to be roughly comparable. And individuals within groups vary considerably, depending on their environmental backgrounds and the influence of specific genes.

Perhaps the best example of this comes from the most obvious genetic difference between two populations: sex. Nobody denies that men are, on the average, physically stronger than women. Nobody disagrees that while strength is a complex trait, it is powerfully influenced by genes. There are individual women who are stronger than many men and individual men who are weaker than most women. But men are generally stronger than women; and the difference has little to do with the environment.

Nevertheless, to create a stereotype of male superiority on the basis of this difference would be a mistake. The genes actually provide compensation for the smaller muscles and frame of the female physique. In the world of gymnastics, for instance, the various events emphasize the characteristics in which each sex excels. The men's events—the parallel bars, rings, side horse—place a premium on strength and agility. The women's—balance beam, uneven parallel bars—demand grace, balance, and flexibility. Even the events that the two sexes perform in common, the vault and floor exercise, are approached and evaluated differently, according to sex. In gymnastics, at least, value judgments about the relative merits of the physical prowess of each sex are meaningless.

In health, the differences are even more astonishing. By now there is general medical agreement that women have a decided advantage over men in fighting disease. The difference exists from the moment of conception until puberty. Males are spontaneously aborted more frequently, are more susceptible to bacterial and viral infections, and have lower survival rates for chronic diseases such as leukemia than females.

The reasons for this disparity are probably rooted in the genes. One theory holds that because a woman carries two X chromosomes while a man carries only one, and because many of the genes that program the immune system are located on the X chromosome, she may produce more of these immunological weapons than he. If one of her X chromosomes happens to be defective, the other can take up the slack; he has only one shot at good health.

Does this mean that women are genetically superior to men in terms of health? Not at all. In fact, the genes provide their own compensatory mechanism to make up for the male deficiency. While men might be more prone to illness and more susceptible to defects of the X chromosome, sperm carrying the male Y chromosome manage to fertilize about 120 ova for every 100 fertilized by sperm carrying the female X chromosome. And while more male embryos and fetuses are aborted spontaneously, there are still 105 males born for every 100 females. Because boys remain more susceptible to disease than girls during early childhood, the ratio of male to female finally evens out around puberty. Thereafter, the differences in life spans between men and women are due primarily to environmental factors. Even now, men drink more, smoke more, and are more prone to automobile and industrial accidents than women.

In an ideal world, there would be no question as to how we might use this kind of information. The value of genetic prophecy lies in its ability to help us avoid disease and not in its questionable links to value judgments. We would recognize the importance of the genes as a predictive tool and act accordingly.

But this is not an ideal world. The danger of stretching genetic information to support social goals exists. When we see a child who has Down's syndrome, we tend to classify the child as "a mongoloid"—in other words, as subhuman. When we learn that someone's genes have expressed themselves in an unhealthy way, we often blame the victim. And when we discuss ways to deal with an unhealthy work environment, we divide into two extreme points of view: one which places full responsibility on industry to make its processes safe for *all* workers (something that is often technically and financially impossible); and one which views the susceptible worker as the hazard, because, if he or she isn't around, the environment becomes "safe"

In industry, we must search for a balance that will ensure a safe environment for as many workers as possible, while making sure that workers who are at greatest risk either find another job or accept the responsibility for the illness to which they are predisposed. Stigmatizing individuals and groups because of inherited characteristics is rooted in a misunderstanding of the meaning and role of genetics. It is a problem that must be overcome if genetic prophecy is to reach its full potential.

Coping with the Genetic Burden

In 1973, Michael Swift of the University of North Carolina was studying the incidence of cancer among relatives of people with ataxia telangiectasia (AT), a rare and often fatal disease. He knew that AT predisposed its victims to an extremely high risk of cancer. But he found that their relatives also had a higher risk: They were more than five times as likely as the general population to die of cancer by age 45.

Swift's main focus was on the parents of AT victims. Since AT is a recessive genetic disease (which requires that those who contract it carry two AT genes), it stood to reason that each parent of an AT victim was an AT carrier. It became clear, in fact, that those relatives who were at greater risk for cancer carried one AT gene, even though no test yet exists to positively identify people who have that gene.

Swift was not studying some esoteric phenomenon. While AT itself might be rare, estimates have placed the number of single-gene carriers in the United States at about 1 percent of the population, or about two million people. Because the gene has been linked to eight common cancers—leukemia, gastric cancer, and cancers of the breast, ovaries, colon, cervix, gallbladder, and lymph nodes—it was possible that people with the AT gene could account for many of the early cancer deaths in this country.

Swift considered the problems involved in telling the 40 parents of the patients he studied about his results. But he realized he had little choice. He had promised to discuss his findings with them; and he felt that if he could warn them of their propensity for cancer, perhaps they and their physicians would be on the alert for the early signs that could enhance their chances for a cure. He knew that he would be giving them some terrible news. But if he presented it correctly, if he counseled them, keeping in mind both the facts of the situation and their probable reaction, Swift thought that the information he had to offer would prove beneficial.

One by one, Swift saw the parents of the AT victims, taking care to deal with them in person. He carefully outlined the situation, telling them that the odds of their other children carrying one AT gene stood at about two in three, and allowing them to make the decision as to whether or not their own relatives, who could not be confirmed or rejected as AT carriers, should be told. Then he promised to stay in touch with them to answer their questions.

Over the years, Swift's prophecy proved true. The parents he talked to showed an unusually high incidence of cancer. But, the information he had given them seemed to help. Many informed their family physicians and began to submit to regular physical check-ups; one woman was alert enough to bring her doctor's attention to what turned out to be a premalignant tumor of the uterus.

Michael Swift's work cuts to the heart of another controversy surrounding genetic screening. Some investigators contend that the burden of knowing that you are predisposed to a particular disease is debilitating, and that learning that your child's disability is genetic (and has been given directly to him by you) causes immense psychological pain. Some surveys have indicated that this is true. A study in England around 1970 implied that divorces among couples who underwent genetic screening and counseling were three times that of the general population. Other surveys have found that people tend to act out the role of the population of Orchemenos, misinterpreting information that they are given to the point that some people leave a counseling session convinced that they are susceptible to a disease when the counselor has told them specifically that they are not. And still other reseachers have found that people who discover they are carriers of genetic traits that may cause illness in them or their children develop a powerful sense of self-hatred and contempt. Perhaps it is because they view the disease as part of their very fiber, a parasitic slice of themselves.

But Michael Swift's study indicates that this may be true only when the counseling is hurried or careless or insensitive to the personality of the individual being counseled. As we now know, genes by themselves do not cause disease; they require an environmental situation which makes that disease possible. So while genetic problems may seem overwhelming to some people, their attitudes usually stem from inadequate genetic education, an inability on the part of the counselor to explain the sitution fully and carefully, a terrible misinterpretation of the facts, or a need to override what they are being taught. With the right kind of education and counseling, they tend to respond as we would expect them to: constructively, actively, in defense of their health.

The idea that genetic screening will dehumanize medicine, depriving people of the care and protection of their personal physicians is, in this light, absurd. Mass screening does relieve the physician of the task of performing some primary diagnostic functions; it offers an impersonal, impartial evaluation that naturally lacks the warmth of a bedside manner.

But that only means that the physician's role in medicine will begin to change. Because he will pay more attention to counseling his patients about the results of their screening tests, he will still be involved, but as one who helps prevent disease, rather than one who is merely called in to cure it. As Michael Swift pointed out after he accepted the part-time role of counselor: "Hurrying through the process of genetic counseling can cause as many problems as it solves. Counseling requires both time and a firm rapport with the patient. It is a combination of scientific knowledge and humane skills."

The personal side of medicine—the relationship between a doctor and his patients—will not disappear. It will simply undergo a metamorphosis. The physician who acts as a mystical healer today will impart knowledge and understanding tomorrow. And that is what medicine should be all about.

The Price of Success

Clearly, the most important element in avoiding the perils of prophecy is education. Both physicians and laymen have to begin to think genetically and to recognize both the power and the limitations of the genes' influence on our lives.

But are we ready to do so? Some people think that we are not.

In 1974, Edwin Naylor of the State University of New York at Buffalo decided to find out just how knowledgeable physicians and family planning professionals were about the basic facts of genetics. He chose two groups: the membership of a professional society of obstetricians and gynecologists in Pittsburgh, Pennsylvania, and the staff of a Family Planning Council that received its funding from the Department of Health, Education, and Welfare. Both groups were deeply involved in counseling people who want family planning information. Naylor reasoned that they should be among the most informed professionals in their field.

The survey consisted mainly of seven questions designed to measure knowledge of basic genetic principles. The survey asked:

1. What is the basic unit of heredity?
2. What do you feel mental retardation is primarily due to?
3. What is sickle-cell anemia the result of?
4. What is Down's syndrome the result of?
5. At what age does a woman have the greatest risk of having a Down's syndrome child?
6. Which condition cannot be detected by prenatal diagnosis?
7. What is the recurrence risk of PKU [the risk that the other children of two PKU carriers will also contract the disease]?*

The test was multiple choice. The results were astonishing. Over 20 percent of the physicians and nearly 50 percent of

*Answers; 1. The gene. 2. Combination of heredity and environment. 3. Inherited hemoglobin defect. 4. Chromosome aberration. 5. 40 years and over. 6. Sickle-cell anemia. 7. 25 percent.

the family planning professionals did not know that the basic unit of heredity is the gene, or, for that matter, that Down's syndrome is caused by a chromosomal abnormality. Nobody in either group answered all seven questions correctly. The mean score among physicians was about 4½ correct answers, among the family planning professionals less than 3½. The questions both groups most often answered correctly concerned the age of women who have the greatest risk of giving birth to a baby with Down's syndrome and the cause of sickle-cell anemia, mainly because both groups were most familiar with screening programs for those two problems.

Most of the physicians had completed their formal medical education more than 20 years before, about the time Crick and Watson were unveiling the structure of DNA. But they had not managed to pick up genetic information in the continuing education courses that many physicians take to stay up to date with the latest advances. In 1974 in the Pittsburgh area, knowledge about genetics was distressingly poor. Other groups in other areas were not tested, but their performances would probably have been no better.

Not surprisingly, the deficit does not show up in research circles, where investigators are running hot on the trails of countless genetic clues. It appears mainly among practicing physicians and professionals in public health, those most actively involved in dispensing medical services. Those who practice medicine the most understand genetics the least. And the schools from which they graduate hold to the equation: Schools that emphasize research teach genetics; schools primarily devoted to turning out practitioners often do not. A decade or so ago, that might have made sense, since the genes rarely seemed important in clinical settings. Doctors had too much to learn anyway; why burden them with extraneous information?

But now the balance of that equation is changing. More medical schools are requiring that students learn genetic principles and practices. Research is finding its way into the examining room. The press has caught the genetics bug and is putting it on the front page almost daily. And those who persist in viewing genetics as an esoteric science of molecules and fruit flies are being squeezed out of the mainstream of their profession.

Genetic education is reaching the public as well. Such groups as the Biological Sciences Study Curriculum have pointed out that most people have only one chance to pick up the basic principles of genetics in an organized fashion. They are writing new textbooks and designing new courses so that high schools can dump the earthworms and frogs of old for the more relevant, more practical problems of human genetics, human problems, human disease. The change is gradual, but it is picking up speed. Genetics is leaving the laboratory and hitting the streets.

And just in time. For the stigmas and stereotypes that grew out of the XYY and sickle-cell screening programs have not disappeared. The guilt and sense of failure felt by people who are told they are carriers for potentially dangerous genetic traits crop up constantly. The horror of being informed that they cannot count on producing healthy children—and the feelings of inferiority that it causes—still strikes many prospective parents. Unless counseling and education can instill a true understanding of what the results of prophecy mean, we will find ourselves creating a mythology around the impact of the genes, a sense of good and bad that parallels the medical stigmas that once sprung up around cancer, tuberculosis, and other diseases. Unless we begin to counteract ignorance now, we will face its consequences in the future.

Massive education may be difficult, but it is not impossible. Before 1957, public understanding of the potential of space flight seemed as remote as a trip to the moon. Within months after the Soviet Union launched Sputnik, however, the American educational system began to gear up to meet the challenge. Interest in the physical sciences and engineering soared. By 1960, the average citizen knew the names of the first seven astronauts as well as he did his own. Overnight, our attitudes changed. The injection of a political reality into our lives launched a revolution in education.

Genetics is on the verge of offering us a similar jolt. And while the revolution is more gradual, the trend is just as certain. Steeping the public in genetic lore will not negate all the perils of prophecy; but it will alleviate many. Conceiving of genetic variation as inherently threatening, instead of as a natural product of evolution linked inextricably to changes in environment, is largely caused by a lack of knowledge and understanding. Eradicating ignorance is one way of relieving that genetic burden.

Of Rocks and Hard Places

The new genetic knowledge offers us a new set of remedies. Some of them pose the most difficult ethical questions we, as a species, have ever had to answer.

The issue of abortion is already explosive. Genetic prophecy promises to make the situation even more unstable, by making it possible to characterize the health and the genetic potential of the fetus, through amniocentesis (the analysis of the components of the amniotic fluid in which the baby floats), and even through a blood test of the fetus itself, a procedure that is still experimental.

When the testing reveals a healthy child, it can relieve the parents of a small burden of fear. It replaces the time-honored line of obstetricians when they are asked if the child they have just delivered is normal: "It has five fingers on each hand and five toes on each foot."

But when prenatal testing reveals a child with abnormalities, tougher choices emerge. Aborting a child who has a disease as deadly as Tay-Sachs or who has physical defects that make life practically impossible poses a clear-cut choice for most people. Aborting a child with Down's syndrome, which may not threaten life but makes an abnormal existence inevitable, is slightly more troublesome. Aborting a child with the two genes of sickle-cell disease, which is almost always debilitating but occasionally leaves a child practically unaffected, is more difficult still. But what about a child who carries the gene for Huntington's disease, which inevitably strikes around the age of 40 but has no adverse effects on life until then, and may even be treatable by the time that child reaches the crucial age? What of a child who has a predisposition to cancer, but who may live a healthy, long life if the narrow range of environments within which he or she is safe is carefully maintained? What about a child predisposed to manic depression? Or a slightly lower intelligence? Or obesity?

What, in short, constitutes normality?

When genes are as accessible as items on the supermarket shelf, when parents-to-be can choose or reject a child the way they decide on a style of furniture, how will we respond to our new responsibilities? Today, we have few answers. Amniocentesis discloses the unborn baby's sex, and there have been rare instances of couples deciding to abort a child of one sex (usually female) because they wanted a child of the other (usually male). But the overwhelming majority seems prepared to accept either possibility.

Sex, at least, is not a trait that drives most people to the brink of difficult genetic decisions. But other traits might. There is always the chance that physicians, counselors, even governments may bring subtle pressures to bear on women who are carrying children with characteristics considered defective. That it might happen in the case of problems like sickle-cell anemia is bad enough. But what if the pressures influence a mother to abort a child with traits that happen to have fallen out of favor? Are we then on the verge of creating a new eugenics movement, dedicated to wiping out examples of "inferior" breeding in an attempt to protect society?

If we remember the misguided efforts of past eugenics movements, chances are that we will avoid that particular problem. In the past, at least, such movements seem to have run into difficulties in deciding exactly what an inferior trait is. As James Bowman has noted:

> Epileptics Dostoevsky and Julius Caesar, drug users Poe and Rimbaud, psychotics Newton and Van Gogh, blind Milton, deaf and son-of-an-alcoholic Beethoven, crippled Kaiser Wilhelm II and Byron, pauper Mozart, tubercular Schubert, Chopin, and Robert Louis Stevenson, syphilitic and leprous Gauguin, deformed Toulouse-Lautrec and many others would have been classified among the undesirables according to the 1925 Eugenics Society.

As long as abortion for any reason is legal until well into the second trimester of pregnancy, we will have to accept the personal ethics of the parents-to-be as the criteria upon which the decision to abort is based. In some cases, the decision may be more than they can stand: What would you do, for instance, if you were carrying twins and learned that one was severely defective while the other was fine? "Selective abortion for only one twin has actually been achieved. But it may well endanger the healthy fetus even as it terminates the defective one. Would you risk the health of the normal fetus to spare your family, society, and the abnormal child itself the pain and burden of the defect?" Or would you give birth to both, accepting the difficulties of the defective child for the sake of the life of the healthy one?

Beyond the terrible choices of abortion lie the problems of screening the living. Genetic prophecy will undoubtedly increase the amount of information we have that we can do nothing about. Some people contend that screening carries with it the right of the patient to know the results and the responsibility of the tester to inform. That may seem rational enough. But if a predictive test were developed for Huntington's disease, how would we use it? Every child of a parent with Huntington's has a 50 percent chance of coming down with the disease, depending on whether or not he or she inherits the gene. It would be easy enough to inform those who did not inherit the gene; by doing so, we could ease the fear and uncertainty of an unknown genetic burden. But what of those who carry it? Knowing that they are doomed to suffer neurological decline and death soon after the age of 40 must have an inevitable impact on their lives. Some may respond well to the information, others poorly. Some may decide to take the chance of having children, others may go childless. The usefulness of the knowledge depends upon very personal factors. How do we distinguish among those whom it would help and those whom it would hurt?

These are not the only unanswerable questions that will surface as we apply our growing knowledge of the gene. But they are a foretaste of what is to come. We will have the luxury of making choices that have never existed before; but we will also carry a burden of responsibility.

Manipulating Genes

Thus far, genetic prophecy gives us the opportunity to begin a search for some mythical level of genetic perfection. The quest, of course, is a false one. We shall never control all the elements that collaborate to create characteristics. We shall never manage to hold the environment stable long enough to adapt to it ideally; for the process of adaptation itself changes the combination of environmental factors that are needed for perfection.

Nevertheless, prenatal diagnosis and an awareness of our susceptibilities to various environmental factors does put us on the road to improvement. And it won't be long before we are not only controlling environments but genes as well, before we are playing an active role in modifying both sides of the equation. New advances in genetic research are bringing us to the point where we may be able to orchestrate at least some of the genetic heritages of our offspring:

- In the summer of 1980, UCLA Medical Center announced the first animal gene transplant. Scientists treated the bone marrow cells from mice that were genetically sensitive to a certain drug with DNA from the cells of mice that were resistant. When they reintroduced those modified cells into the susceptible mice, the mice became resistant to the drug. Scientists and the press alike boldly predicted that a similar attempt would be made in man within five years.

- Only three months later, two patients suffering from thalassemia (a recessive genetic disease)—one in Israel and one in Italy—were treated in similar fashion by Martin Cline of UCLA. The results of the experiment were inconclusive; Cline resigned during the ensuing controversy. But the attempt to direct a genetic change in humans had taken place.

- In mid-1980, scientists at Yale University succeeded in infecting mouse embryo cells with viruses that carried modified DNA. The genes that they had attached to the viruses became incorporated into the mouse cells; when they developed into mature mice, they carried with them the transplanted characteristics.

- In early 1981, Karl Ilmensee and his colleagues in Geneva, Switzerland reported that they had cloned mice—the first time that a mammal had successfully undergone the procedure. They had transferred nuclei from the body cells of mouse embryo into several fertilized eggs from which they had removed the original nuclei. Then they placed the altered eggs in the uteri of mice with significantly different genetic structures. The eggs grew, taking directions from the blueprints provided by the transplanted nuclei. They developed into three mice, each genetically identical to the original mouse embryo.

Gene alteration, cloning, and *in vitro* fertilization—the technique of developing so-called test-tube babies—place us on the verge of actively intervening in the process of inheritance. As they are refined, the three techniques will allow us to move genes around at will, as if they were tiny parts of a vast and complex Tinker Toy. Once we discover what a gene does, we can snip it out of its rightful place in a length of DNA and introduce it into any other genetic chain. And we can perform this operation either with a natural gene from a cell or with genetic material produced from chemicals on the laboratory shelf.

As our sophistication has increased, our goals have expanded. In the mid-1970s, bacterial and viral genes were the only ones moved from cell to cell; by the end of the decade, human genes, such as the one for insulin, were transplanted into bacteria. The next obvious step is to begin to shuttle genes between animals higher on the evolutionary ladder, up to and including man.

There are two targets for human genetic alteration. The first includes adults who may be carrying a clearly defective gene or groups of genes. For them, the critical alterations will be aimed at changing a specific genetic response to the environment. For diabetics, the ability to produce insulin may be delivered to the pancreas; for those with sickle-cell anemia, the bone marrow, which produces the defective hemoglobin, may be replaced. Genetic therapy in adults will home in on the particular organ affected (and only that organ), and the resulting change will not be passed on to their children. It is for that reason that fewer ethical issues will arise from adult gene therapy. In essence, it is no different from any other organ transplant.

The second set of targets, the fertilized egg and the sex cells (sperm and egg), poses far greater problems. If an egg that is recognized as defective is altered, the change will affect *all* cells descended from that egg, including the sex cells. The change will be passed on through the generations, and the natural integrity of a human genetic line will have been permanently ruptured. Whether or not that has terrible consequences remains to be seen. But the mere fact that we can control and direct human reproduction, both by removing genes that we consider deleterious and by recognizing and encouraging the development of sperm and egg that already have characteristics we consider desirable, means that we are gaining control over the evolution of the human gene pool—the rich and extraordinary plexus of variation that has allowed us to develop as we have.

Despite the fact that most experts do not expect to see these techniques used routinely (if at all) until at least the twenty-first century, it is not too early to start asking the right questions. Already, the President's Commission on Biomedical Ethics and Research has begun to examine the issues; already, special interest groups are gearing up to fight the aspects of the genetic revolution that they view as particularly dangerous; already, the first incursions into the world of directed genetic manipulation have taken place. Using genes to foretell the future is only one of many ways that genetics is finding its way into our lives, but it is one of the more visible. As it grows in sophistication and accuracy, it will become more and more indispensible to our medicine, our professions, the way we run our lives. As such, we must use it fully, but we must make sure that it does not begin to use us. Genetic prophecy is still at the tail of society's whip, but it is moving inexorably toward the handle.

APPENDIX

Screening Centers

The following laboratories are recognized by the National Institutes of Health as qualified to conduct screening for genetic disorders. Many of them are capable of performing any of the tests described in this book, although the tests are not all performed routinely. Some, of course, have more limited facilities. Laboratories that are more likely to perform the less routine tests are indicated by an asterisk.

ALABAMA

Birmingham

University of Alabama in Birmingham
The Medical Center
University Station
1720 Seventh Avenue South
Birmingham, Alabama 35294

Mobile

University of Southern Alabama Medical Center School of Medicine
Moorer Building
2451 Fillingim Street
Mobile, Alabama 36617

ALASKA

Fairbanks

WAMI Medical Education Program
University of Alaska
Arctic Health Research Building
Fairbanks, Alaska 99701

ARIZONA

Phoenix

Genetics Center of the Southwest Biomedical Research Institute
123 East University Drive
Tempe, Arizona 85281

Source: *Clinical Genetic Service Centers: A National Listing,* National Clearing House for Genetic Diseases, DHHS Publication No. (HSA) 80–5135.

St. Joseph's Hospital and
Medical Center
350 West Thomas Road
P.O. Box 2071
Phoenix, Arizona   85003

Tucson

University of Arizona Health
Sciences Center
College of Medicine
1501 North Campbell Avenue
Tuscon, Arizona   85724

ARKANSAS

Little Rock

University of Arkansas
College of Medicine
4301 West Markham
Little Rock, Arkansas   72201

CALIFORNIA

* **Duarte**

City of Hope National Medical Center
1500 East Duarte Road
Duarte, California   91010

Fresno

Valley Children's Hospital
3151 North Millbrook
Fresno, California   93703

Loma Linda

Loma Linda University Medical Center
11234 Anderson Street
Loma Linda, California
92354

Los Angeles

Children's Hospital of Los
Angeles
4650 Sunset Boulevard
Los Angeles, California
90027

* Los Angeles County—Uni-
versity of Southern California Medical Center
General Lab Building
1129 North State Street
Los Angeles, California
90033

Los Angeles County—Uni-
versity of Southern California Medical Center
1200 North State Street
Los Angeles, California
90033

Martin Luther King, Jr., Gen-
eral Hospital
Charles R. Drew Postgraduate
Medical School
12021 South Wilmington
Avenue
Los Angeles, California
90059

* University of California Los
Angeles
Center for Health Services
760 Westwood Plaza
Los Angeles, California
90024

Oakland

Children's Hospital Medical
Center of Northern
California
51st and Grove Streets
Oakland, California   94609

Orange

University of California, Irvine
Medical Center
101 City Drive South
Orange County, California
92668

Sacramento

University of California, Davis
School of Medicine
4301 X Street
Sacramento, California
95817

San Diego

Children's Hospital and
Health Center San Diego
8001 Frost Street
San Diego, California 92123

University of California, San
Diego Medical Center
University Hospital
225 West Dickenson Street
San Diego, California 92103

San Francisco

* University of California San
Francisco
3rd and Parnassus Avenue
San Francisco, California
94143

Stanford

* Stanford University Medical
Center
300 Pasteur Drive
Stanford, California 94305

Torrance

* Harbor UCLA Medical
Center
1000 West Carson Street
Torrance, California 90509

COLORADO

Denver

* University of Colorado Health
Sciences Center
4200 East Ninth Avenue
Denver, Colorado 80262

* National Jewish Hospital
3800 East Colfax Avenue
Denver, Colorado 80206

CONNECTICUT

Connecticut Department of
Health Services
Genetics Information
Center
79 Elm Street
Hartford, Connecticut
06115

Farmington

University of Connecticut
Health Center
Farmington, Connecticut
06032

New Haven

* Yale University School of
Medicine
333 Cedar Street
New Haven, Connecticut
06510

DELAWARE

Wilmington

A. I. DuPont Institute
P.O. Box 269
Wilmington, Delaware 19899

Wilmington Medical Center
P.O. Box 198999
Wilmington, Delaware 19899

DISTRICT OF COLUMBIA

Children's Hospital National
  Medical Center
111 Michigan Avenue, N. W.
Washington, District of
  Columbia   20010

George Washington Univer-
  sity
2150 Pennsylvania Avenue,
  N. W.
Washington, District of
  Columbia   20037

Georgetown University Medical Center
3800 Reservoir Road, N. W.
Washington, District of
  Columbia   20007

Howard University
College of Medicine
Box 75
520 W Street, N.W.
Washington, District of
  Columbia   20001

FLORIDA

Gainesville

University of Florida
College of Medicine
J. Hillis Miller Health Center
Gainesville, Florida   32610

Miami

University of Miami
School of Medicine
1601 Northwest 12th Avenue
P. O. Box 016820
Miami, Florida   33101

Tampa

University of South Florida
College of Medicine
Box 15
12901 North 30th Street
Tampa, Florida   33612

GEORGIA

Atlanta

Emory University School of
  Medicine
Box AM
Atlanta, Georgia   30322

Georgia Mental Health Insti-
  tute
Emory University School of
  Medicine
1256 Briarcliff Road
Atlanta, Georgia   30306

Augusta

Medical College of Georgia
1120 Fifteen Street
Augusta, Georgia   30912

HAWAII

Honolulu

University of Hawaii
Kapiolani Children's Medical
  Center
1310 Punahou Street
Honolulu, Hawaii   96826

IDAHO

Boise

Idaho Department of Health and Welfare
2220 Old Penitentiary Road
Boise, Idaho 83702

ILLINOIS

Chicago

Children's Memorial Hospital
* Northwestern University Medical School
2300 Children's Plaza
Chicago, Illinois 60614

Cook County Hospital
Children's Hospital
700 South Wood Street
Chicago, Illinois 60612

Michael Reese Hospital and Medical Center
29th Street and Ellis Avenue
Chicago, Illinois 60616

* Prentice Women's Hospital
Northwestern University Medical School
333 East Superior Street
Chicago, Illinois 60611

Rush-Presbyterian-St. Luke's Medical Center
1753 West Congress Parkway
Chicago, Illinois 60612

* University of Chicago Hospital
Pritzker School of Medicine
950 East 59th Street
Chicago, Illinois 60637

University of Illinois Medical Center
Abraham Lincoln School of Medicine
840 South Wood Street
Chicago, Illinois 60612

Peoria

University of Illinois College of Medicine
Allied Agencies, Inc.
123 S.W. Glendale Avenue
Peoria, Illinois 61605

Springfield

Southern Illinois University School of Medicine
P. O. Box 3926
Springfield, Illinois 62708

Urbana

Regional Health Resource Center
1408 West University Avenue
Urbana, Illinois 61801

INDIANA

Bluffton

Caylor-Nickel Research Institute
311 South Scott Street
Bluffton, Indiana 46714

Indianapolis

Indiana University School of Medicine
1100 West Michigan Street
Indianapolis, Indiana 46223

Methodist Hospital of Indianapolis
1604 North Capitol Avenue
Indianapolis, Indiana 46206

South Bend

Memorial Hospital
615 North Michigan Street
South Bend, Indiana   46601

IOWA

Iowa City

* University of Iowa Hospital
  and Clinics
Iowa City, Iowa   52242

KANSAS

Kansas City

* Kansas University Hospital
  College of Health Sciences
39th and Rainbow Boulevard
Kansas City, Kansas   66103

Topeka

Lattimore-Fink Laboratories,
  Inc.
115 West Crane Street
Topeka, Kansas   66603

Northeast Kansas Genetic
  Counseling Center
1518 Southwest 8th Street
Topeka, Kansas   66604

Wichita

University of Kansas School
  of Medicine-Wichita
1001 North Minneapolis
  Street
Wichita, Kansas   67214

KENTUCKY

Lexington

* University of Kentucky School
  of Medicine
Medical Center
800 Rose Street
Lexington, Kentucky 40536

Louisville

University of Louisville Medical School
334 East Broadway
Louisville, Kentucky   40202

LOUISIANA

New Orleans

* Louisiana Heritable Disease
  Center
Louisiana State University
  Medical Center
1542 Tulane Avenue
New Orleans, Louisiana
  70112

Tulane University School of
  Medicine
1430 Tulane Avenue
New Orleans, Louisiana
  70112

Shreveport

Louisiana State University
  School of Medicine
University Hospital
1501 Kings Highway
Shreveport, Louisiana
  71130

MAINE

Bangor

* Eastern Maine Medical
  Center
489 State Street
Bangor, Maine   04401

Bar Harbor

* Center for Human Genetics
Firehouse Hill
Bar Harbor, Maine   04609

Portland

Maine Medical Center
22 Bramhall Street
Portland, Maine   04102

Scarborough

Foundation for Blood
   Research
P. O. Box 426
Scarborough, Maine   04074

MARYLAND

Baltimore

John F. Kennedy Institute
707 North Broadway
Baltimore, Maryland   21205

* Johns Hopkins University
   School of Medicine
601 North Broadway
Baltimore, Maryland   21205

University of Maryland Hospital
22 South Greene Street
Baltimore, Maryland   21201

Bethesda

* National Institutes of Health
Building 10, Room 1D21
Bethesda, Maryland   20205

Towson

Greater Baltimore Medical
   Center
6701 North Charles Street
Towson, Maryland   21204

MASSACHUSETTS

Boston

* Boston Hospital for Women
S. G. Mudd Building, Room
   209
250 Longwood Avenue
Boston, Massachusetts
   02115

* Children's Hospital Medical
   Center
300 Longwood Avenue
Boston, Massachusetts
   02115

* Massachusetts General Hospital
Boston, Massachusetts
   02114

* Peter Bent Brigham Hospital
721 Huntington Avenue
Boston, Massachusetts
   02115

* Tufts New England Medical
   Center Hospital
171 Harrison Avenue
Boston, Massachusetts
   02111

Waltham

* Eunice Kennedy Shriver
   Center
200 Trapelo Road
Waltham, Massachusetts
   02154

Worcester

University of Massachusetts
   Medical Center
55 Lake Avenue North
Worcester, Massachusetts
   06105

MICHIGAN

Ann Arbor

* University of Michigan Medical School
1137 East Catherine Street
Ann Arbor, Michigan   48109

Detroit

C. S. Mott Center for Human Growth and Development
Wayne State University School of Medicine
275 East Hancock
Detroit, Michigan   48201

Henry Ford Hospital
2799 West Grand Boulevard
Detroit, Michigan   48202

University of Detroit
4001 W. McNichols Road
Detroit, Michigan   48221

Wayne State University Children's Hospital of Michigan
3901 Beaubien Boulevard
Detroit, Michigan   48201

East Lansing

Michigan State University
B240 Life Sciences Building
East Lansing, Michigan 48824

Grand Rapids

Blodgett Memorial Medical Center
1840 Wealthy Street, S. E.
Grand Rapids, Michigan 49506

Butterworth Hospital
100 Michigan Street, N. E.
Grand Rapids, Michigan 49503

Northville

Michigan Department of Mental Health
18471 Haggerty Road
Northville, Michigan   48167

Royal Oak

William Beaumont Hospital
3601 West 13 Mile Road
Royal Oak, Michigan   48072

MINNESOTA

Minneapolis

Hennepin County Medical Center
701 Park Avenue
Minneapolis, Minnesota 55415

Minnesota State Board of Health
717 S. E. Delaware
Minneapolis, Minnesota 55440

* University of Minnesota Dight Institute for Human Genetics
400 Church Street S. E.
Minneapolis, Minnesota 55455

University of Minnesota School of Dentistry and Health Science Center
Health Science Unit A-16
515 Delaware Street, S. E.
Minneapolis, Minnesota 55455

* University of Minnesota Hospital
Mayo Memorial Building
420 Delaware Street, S. E.
Minneapolis, Minnesota 55455

Rochester

Mayo Clinic
200 First Street, S. W.
Rochester, Minnesota 55901

St. Paul

Gillette Children's Hospital
200 East University Avenue
St. Paul, Minnesota 55101

St. Paul Ramsey Medical Center
640 Jackson Street
St. Paul, Minnesota 55101

MISSISSIPPI

Hattiesburg

University of Southern Mississippi
College of Science and Technology
P. O. Box 8421
Hattiesburg, Mississippi 39401

Jackson

University of Mississippi Medical Center
2500 North State Street
Jackson, Mississippi 39216

Biloxi

Keesler Air Force Base
U. S. Air Force Medical Center
Keesler Air Force Base, Mississippi 39534

MISSOURI

Columbia

University of Missouri Medical Center at Columbia
807 Stadium Road
Columbia, Missouri 65201

Kansas City

Children's Mercy Hospital
University of Missouri at Kansas City
24th and Gillham Road
Kansas City, Missouri 64108

St. Louis

Cardinal Glennon Memorial Hospital for Children
1465 South Grand Boulevard
St. Louis, Missouri 63104

St. Louis Children's Hospital
Washington University Medical School
500 South Kings Highway
St. Louis, Missouri 63110

Washington University Medical School
4911 Barnes Hospital Plaza
St. Louis, Missouri 63110

MONTANA

Helena

Shodair Crippled Children's
Hospital
840 Helena Avenue
Helena, Montana   59601

NEBRASKA

Omaha

* Children's Memorial Hospital
44th and Dewey Avenue
Omaha, Nebraska   68105

St. Joseph's Hospital
Creighton University School
of Medicine
601 North 30th Street
Omaha, Nebraska   68131

University of Nebraska Medical Center
42nd and Dewey Avenue
Omaha, Nebraska   68105

NEW HAMPSHIRE

Hanover

* Dartmouth-Hitchcock Medical Center
Dartmouth Medical School
Hanover, New Hampshire
03755

NEW JERSEY

Camden

* Institute for Medical Research
Copewood Street
Camden, New Jersey   08103

Jersey City

Jersey City Medical Center
50 Baldwin Avenue
Jersey City, New Jersey
07304

Lyons

Veterans Administration Hospital
Lyons, New Jersey   07939

Newark

College of Medicine & Dentistry of New Jersey
New Jersey Medical School
100 Bergen Street, F 532
Newark, New Jersey   07103

Piscataway

College of Medicine & Den-
tistry of New Jersey
Rutgers Medical School
Hoes Lane
Piscataway, New Jersey
08854

NEW MEXICO

Albuquerque

University of New Mexico Medical Center
2211 Lomas NE
Albuquerque, New Mexico
87131

NEW YORK

Albany

* Albany Medical College
New Scotland Avenue
Albany, New York   12208

* New York State Health De-
partment
Empire State Plaza
Albany, New York   12201

Buffalo

Buffalo General Hospital
State University of New York
at Buffalo, School of Medicine
100 High Street
Buffalo, New York   14203

* Children's Hospital of Buffalo
State University of New York
at Buffalo, School of Medicine
86 Hodge Avenue
Buffalo, New York   14222

Manhasset

* North Shore University Hospital
Cornell University College of
Medicine
300 Community Drive
Manhasset, New York
11030

New Hyde Park

* Long Island Jewish-Hillside
Medical Center
270-05 76th Avenue
New Hyde Park, New York
11042

New York City

* Albert Einstein College of
Medicine
Building U, Room 517
1300 Morris Park Avenue
Bronx, New York   10461

* Albert Einstein College of
Medicine
Rose F. Kennedy Center
1410 Pelham Parkway
Bronx, New York   10461

Brookdale Hospital Center
Linden Boulevard at Brook-
dale Plaza
Brooklyn, New York   11212

Brooklyn Hospital
Brooklyn-Cumberland Medi-
cal Center
121 DeKalb Avenue
Brooklyn, New York   11201

* Columbia-Presbyterian Medical Center
622 West 168 Street
New York, New York   10032

Long Island College Hospital
340 Henry Street
Brooklyn, New York   11201

Long Island Jewish-Hillside
Medical Center
Queens Hospital Center
Building J
82-68 164th Street
Jamaica, New York   11432

* Mt. Sinai School of Medicine
Fifth Avenue and 100th Street
New York, New York   10029

* New York Hospital
Cornell University Medical
Center
525 East 68th Street
New York, New York   10021

New York Medical College
Lincoln Hospital
234 East 149 Street
Bronx, New York   10454

New York Medical College
Metropolitan Hospital
1901 First Avenue
New York, New York   10029

New York University Medical
   Center
Room MSB-136
550 First Avenue
New York, New York   10016

* Prenatal Diagnosis Laboratory of New York City
Medical and Health Research
   Association
455 First Avenue
New York, New York
   10016

State University of New York
   Downstate Medical Center
450 Clarkson Avenue
Brooklyn, New York   11203

Rochester

Strong Memorial Hospital
University of Rochester
   School of Medicine
601 Elmwood Avenue
Rochester, New York   14642

Stony Brook

State University of New York
   at Stony Brook
Health Sciences Center T-9
Stony Brook, New York
   11794

Syracuse

* State University of New York
Upstate Medical Center
750 East Adams Street
Syracuse, New York   13210

Thiells

Letchworth Village Devel-
   opmental Center
Thiells, New York   10984

Valhalla

Westchester County Medical
   Center
Valhalla, New York   10595

NORTH CAROLINA

Chapel Hill

* University of North Carolina
   at Chapel Hill
School of Medicine
Chapel Hill, North Carolina
   27514

Charlotte

Charlotte Memorial Hospital
   & Medical Center
1000 Blythe Boulevard
P. O. Box 32861
Charlotte, North Carolina
   28232

Durham

* Duke University Medical
   Center
Durham, North Carolina
   27710

Greensboro

Moses H. Cone Memorial
   Hospital
1200 North Elm Street
Greensboro, North Carolina
   27420

Greenville

East Carolina University
School of Medicine
Greenville, North Carolina
27834

Winston-Salem

Bowman Gray School of Medicine
300 South Hawthorne Road
Winston-Salem, North Carolina 27103

NORTH DAKOTA

Grand Forks

University of North Dakota
School of Medicine
Grand Forks, North Dakota
58202

OHIO

Akron

The Children's Hospital
Medical Center of Akron
281 Locust Street
Akron, Ohio 44308

Cincinnati

Children's Hospital Medical
Center
Pavilion Building
Elland and Bethesda Avenues
Cincinnati, Ohio 45229

Cleveland

Case Western Reserve University
School of Medicine
2119 Abington Road
Cleveland, Ohio 44106

Cleveland Metropolitan
General Hospital
3395 Scranton Road
Cleveland, Ohio 44109

University Hospitals
2065 Adelbert Road
Cleveland, Ohio 44106

Columbus

* Children's Hospital
700 Children's Drive
Columbus, Ohio 43205

Dayton

Children's Medical Center
1735 Chapel Street
Dayton, Ohio 45404

St. Elizabeth Medical Center
601 Miami Boulevard West
Dayton, Ohio 45408

Kettering

Kettering Medical Center
3535 Southern Boulevard
Kettering, Ohio 45429

Mount Vernon

Mount Vernon Developmental
Center
Mount Vernon, Ohio 43050

Toledo

Medical College of Ohio at
Toledo
Caller Service Number: 10008
Toledo, Ohio 43699

OKLAHOMA

Oklahoma City

Children's Memorial Hospital
940 N. E. 13th Street
P. O. Box 26901
Oklahoma City, Oklahoma
  73190

University Hospital
University of Oklahoma
  Health Service Center
800 N. E. 13th Street
Oklahoma City, Oklahoma
  73190

Tulsa

Children's Medical Center
5300 East Skelly Drive
Tulsa, Oklahoma   74135

OREGON

Portland

* University of Oregon Health
  Sciences Center
P. O. Box 574
Portland, Oregon   97207

PENNSYLVANIA

Danville

Geisinger Medical Center
North Academy Avenue
Danville, Pennsylvania
  17821

Hershey

* Milton S. Hershey Medical
  Center
Pennsylvania State University College of Medicine
500 University Drive
Hershey, Pennsylvania
  17033

Philadelphia

Hahnemann Medical College
  and Hospital
230 North Broad Street
Philadelphia, Pennsylvania
  19102

Jefferson Medical College
Room 710 College
1025 Walnut Street
Philadelphia, Pennsylvania
  19107

St. Christopher's Hospital for
  Children
2600 North Lawrence Street
Philadelphia, Pennsylvania
  19133

Temple University Medical
  School
3400 North Broad Street
Philadelphia, Pennsylvania
  19122

University of Pennsylvania
Children's Hospital of Philadelphia
34th and Civic Center Boulevard
Philadelphia, Pennsylvania
  19104

University of Pennsylvania
University Hospital
3400 Spruce Street
Philadelphia, Pennsylvania
  19104

Pittsburgh

Presbyterian-University Hospital
270 DeSoto Street
Pittsburgh, Pennsylvania
  15213

University of Pittsburgh
Children's Hospital
125 DeSoto Street
Pittsburgh, Pennsylvania
15213

University of Pittsburgh
Magee-Women's Hospital
Forbes and Halkett Avenues
Pittsburgh, Pennsylvania
15213

RHODE ISLAND

Providence

Rhode Island Hospital
593 Eddy Street
Providence, Rhode Island
02902

Women and Infants Hospital of Rhode Island
50 Maude Street
Providence, Rhode Island
02908

SOUTH CAROLINA

Charleston

Medical University of South
Carolina
171 Ashley Avenue
Charleston, South Carolina
29403

South Carolina Department
of Mental Retardation
Coastal Center
41 Bee Street
Charleston, South Carolina
29401

Columbia

South Carolina Department of
Mental Health
P.O. Box 119
Columbia, South Carolina
29202

University of South Carolina
School of Medicine
3321 Medical Park Road,
Suite 302
Columbia, South Carolina
29203

Greenwood

Greenwood Genetic Center
1020 Spring Street at Ellenberg
Greenwood, South Carolina
29646

SOUTH DAKOTA

Vermillion

University of South Dakota
School of Medicine
Clark and Dakota
Vermillion, South Dakota
57069

TENNESSEE

Johnson City

East Tennessee State College
of Medicine
P. O. Box 19840-A
Johnson City, Tennessee
37601

Knoxville

University of Tennessee
Memorial Research Center &
Hospital
1924 Alcoa Highway
Knoxville, Tennessee   37920

Memphis

University of Tennessee
Center for Health Sciences
711 Jefferson Avenue
Memphis, Tennessee   38163

Nashville

Meharry Medical College
1005 18th Avenue
Nashville, Tennessee   37208

Vanderbilt University School of
    Medicine
Nashville, Tennessee   37232

TEXAS

Dallas

* University of Texas Health
    Science Center at Dallas
Southwestern Medical School
5323 Harry Hines Boulevard
Dallas, Texas   75235

Denton

Texas Department of Mental Health and Mental Retardation
404 West Oak Street
Denton, Texas   76201

El Paso

El Paso Rehabilitation Center
2630 Richmond Street
El Paso, Texas   79930

Galveston

University of Texas Medical
    Branch at Galveston
Child Health Center
Galveston, Texas   77550

Houston

Baylor College of Medicine
Texas Children's Hospital
6621 Fannin
Houston, Texas   77030

* M. D. Anderson Hospital and
    Tumor Institute
Texas Medical Center
Houston, Texas   77030

University of Texas Medical
    School of Houston
Texas Medical Center
P. O. Box 20708
Houston, Texas   77025

San Antonio

Santa Rosa Medical
    Center
P. O. Box 7330, Station A
San Antonio, Texas   78285

University of Texas Health
    Science Center at San
    Antonio
7703 Floyd Curl Drive
San Antonio, Texas   78284

UTAH

American Fork

Utah State Training School
American Fork, Utah   84003

Salt Lake City

University of Utah Medical
    Center
50 North Medical Drive
Salt Lake City, Utah   84132

VERMONT

Burlington

* University of Vermont
College of Medicine
A115 Medical Alumni Building
Burlington, Vermont   05401

VIRGINIA

Charlottesville

University of Virginia School
of Medicine
Charlottesville, Virginia
22908

Norfolk

DePaul Hospital
150 Kingsley Lane
Norfolk, Virginia   23505

Eastern Virginia Medical
School
P. O. Box 1980
Norfolk, Virginia   23501

Richmond

* Medical College of Virginia
11th and Marshall
Box 33, MCV Station
Richmond, Virginia   23298

WASHINGTON

Seattle

* Center for Inherited Diseases
University of Washington
University Hospital
1959 East Pacific Street
Seattle, Washington   98195

Children's Orthopedic Hospital
4800 Sand Point Way, N. E.
Seattle, Washington   98105

Department of Social and
Health Services
Child Health Section
1704 N. E. 150th Street
Seattle, Washington   98155

* Veteran's Administration
Medical Center
4435 Beacon Avenue South
Seattle, Washington   98108

Spokane

Inland Empire Genetics
Counseling Service
West 800 Fifth Avenue
Spokane, Washington   99210

Tacoma

Mary Bridge Children's Hospital
311 L Street
Tacoma, Washington   98405

Walla Walla

Saint Mary Community Hos-
pital
401 West Poplar
Walla Walla, Washington
99362

WEST VIRGINIA

Huntington

Marshall University
Huntington, West Virginia
25701

Morgantown

West Virginia State and Re-
gional Genetics Center
West Virginia University
Medical Center
Morgantown, West Virginia
26506

WISCONSIN

Madison

Central Wisconsin Center
317 Knutson Drive
Madison, Wisconsin   53704

* University of Wisconsin at
    Madison
  328 Waisman Center
  1500 Highland Avenue
  Madison, Wisconsin   53706

Marshfield

Marshfield Clinic
1000 North Oak Street
Marshfield, Wisconsin
  54449

Milwaukee

Medical College of Wisconsin
  County General Hospital
8700 West Wisconsin Avenue
Milwaukee, Wisconsin
  53226

Milwaukee Children's Hospital
1700 West Wisconsin Avenue
Milwaukee, Wisconsin
  53233

University of Wisconsin
  Medical School, Madison
Milwaukee Clinical Campus
Mount Sinai Medical Center
950 North 12th Street
Milwaukee, Wisconsin
  53201

SOURCE NOTES

Chapter 1

P. 7. MOTULSKY SOUGHT OUT The story was pieced together from a variety of sources: Marcello Siniscalco, "Field and Laboratory Studies on Favism and Thalassemia in Sardinia," in E. Goldschmidt, ed., *The Genetics of Migrant and Isolate Populations*, Baltimore, Williams & Wilkins, 1963; Dan Keller, *G-6-PD Deficiency*, Cleveland, CRC Press, 1973; A. G. Motulsky et al., "Population Distribution of Inherited Red Cell Enzyme (Glucose-6-Phosphate Dehydrogenase) Deficiency," *Proceedings of the VIIIth International Congress of Hemotology*, September 4–10, 1960, Tokyo, Pan Pacific Press; Marcello Siniscalco, personal communication.

P. 14. RECENTLY, CHARLES GLUECK Jean L. Marx, "The HDL: The Good Cholesterol Carriers?" *Science* 205: 677–679 (1979).

P. 14. IN 1980 ALONE *Heart Facts 1981*, American Heart Association, Inc. 1980.

P. 17. ROBERT LEVY, DIRECTOR Jane E. Brody, "Protein that Lowers Heart Risk Tied to Moderate Drinking and Exercise," *New York Times*, November 13, 1980, p. A33; Robert Levy, personal communication.

P. 17. IN 1974, SCIENTISTS AT R. Friedman and J. Iwai, "Genetic Predisposition and Stress-Induced Hypertension," *Science* 193: 161 (1976).

Chapter 2

P. 21. IN WHAT IS Paul Beeson and W. McDermott, *Textbook of Medicine*, 14th ed., Philadelphia, Saunders, p. 185.

P. 22. "VULNERABLE BODIES ARE . . ." Rhazes' *Treatise on the Small Pox and Measles*, trans. Thomas Stack, in Richard Mead, *A Discourse on the Small Pox and Measles*, London: Printed for John Brindley, 1748, pp. 121–131, quoted in G. Marks and W. K. Beatty, *Epidemics*, New York, Scribner's, 1976, p. 55.

P. 23. "DURING THESE TIMES . . ." Procopius, trans. H. B. Dewing, 6 vols., London: Heinemann, 1914, I:451, quoted in *ibid.*, p. 44.

P. 23. PROCOPIUS WROTE THAT *ibid.*, p. 52.

P. 24. "WHEREVER THEY WANDERED . . ." M. J. Howell,

The Plague at Eyam, "The Practitioner CCII" (1969):100–101, quoted in *ibid.,* p. 140.

P. 25. ACCORDING TO THE   *Surgeon General's Report on Health Promotion and Disease Prevention,* HEW Publication 79-55071, 1979.

P. 26. "BUT ALL OUR EXPERIENCE . . ."   Barton Childs, personal communication.

P. 26. "IT HAS REMAINED . . ."   Charles Scriver, "The William Allan Memorial Address: On Phosphate Transport and Genetic Screening. 'Understanding Backward—Living Forward,' " in Human Genetics, *American Journal of Human Genetics,* 31:243–263 (1979).

P. 27. ACCORDING TO THE   Gerald Stine, *Biosocial Genetics,* New York, Macmillan, 1978, p. 253.

P. 30. "MORE PEOPLE ARE . . ."   Gretchen Kolsrud, personal communication.

P. 32. IN A SURVEY   Barton Childs, personal communication; B. Childs *et al.,* "Human Genetics Teaching in U. S. Medical Schools," paper for publication, April 9, 1980.

P. 36. "INSTEAD OF ASKING . . ."   Barton Childs, personal communication.

Chapter 3

P. 39. ABOUT TWENTY YEARS AGO   Frank Lilly et al., "Genetic Basis of Susceptibility to Viral Leukaemogenesis," *Lancet* 2:1207–1209 (1964); Frank Lilly, personal communication.

P. 44. JEAN DAUSSET COULD STILL   J. Dausset, "Correlation Between Histocompatability Antigens and Susceptibility to Illness," *Progress in Clinical Immunology* 1: 183 (1972).

P. 45. STUDIES HAVE SHOWN   M. A. Khan et al., "Low Frequency of HLA-B27 in American Blacks with Ankylosing Spondylitis," *Clinical Research* 24: 331A (1976).

P. 50. IN 1962, THE CRUISER   H. R. Noer, "An 'Experimental' Epidemic of Reiter's Syndrome," *Journal of the American Medical Association,* 198: 693–698 (1966).

P. 51. THESE FINDINGS WERE   A. Calin and J. Fries, "An 'Experimental' Epidemic of Reiter's Syndrome Revisited: Follow-up Evidence on Genetic and Environmental Factors," *Annals of Internal Medicine,* 84: 564 (1976).

P. 52. ONE SUCH USE   Stanley Hoppenfeld, personal communication.

P. 53. DERMATOLOGISTS HAVE ESTABLISHED   W. Lobitz, "Summary of the Dermatology Workshop of the HLA and Disease Symposium," Paris, June 1976.

P. 53. GASTROENTEROLOGISTS HAVE FOUND   W. E. Braun, *HLA and Disease: A Comprehensive Review,* Cleveland, CRC Press, 1979, pp. 64–70.

P. 53. MYASTHENIA GRAVIS, TO WHICH   R. Pirskanen and

A. Tiilikainen, "Myathenia Gravis and HLA," in *HLA and Disease*, Paris, INSERM, 1976, p. 79.

P. 54. BLADDER CANCER IS "Genetic Association with Bladder Cancer," *British Medical Journal*, 1 September 1979, p. 514.

P. 54. DEPRESSIVE DISORDERS NOT ONLY Lowell Weitkamp et al., "Depressive Disorders and HLA: A Gene on Chromosome 6 that Can Affect Behavior," *The New England Journal of Medicine*, 305: 1301–1306 (1981)."

P. 54. SOME RESEARCHERS CONTEND B. B. Levine et al., "Ragweed Hay Fever: Genetic Control and Linkage to HL-A Haplotypes," *Science* 178: 1201 (1972); D. G. Marsh et al., "Mapping of Postulated Ir Genes within HLA by Studies in Allergic Populations in *HLA and Disease*, 1976, INSERM, Paris, p. 181.

P. 54. SCIENTISTS HAVE FOUND Charles Lange, "Good Genes: Methuselah's Secret," *Chemtech*, February 1979, p. 117.

P. 56. PABLO RUBINSTEIN AND P. Rubinstein et al., "The HLA System in the Families of Patients with Juvenile Diabetes Mellitus," *Journal of Experimental Medicine*, 143: 1277 (1976).

P. 58. HLA-B12 HAS BEEN R. T. D. Oliver et al., "HL-A Associated Resistance Factors and Myelogenous Leukemia," In *HLA and Disease*, Paris, INSERM, 1976, p. 231.

P. 58. PEOPLE WITH HLA-A1 AND HLA-B8 F. Kissmeyer-Nielson et al., "HLA and Hodgkin's Disease: III. A Prospective Study," *Transplantation Review* 22: 168 (1975).

P. 58. AND IN ONE RECENT SURVEY B. Osoba and J. Falk, "Genes May Indicate Prognosis in Lymphoma," cited in Medical News, *Journal of the American Medical Association* 235: 2808 (1979).

P. 58. THE MOST INTERESTING CONNECTION M. J. Giphart and J. D'Amaro, "The Association of HLA-B18 with Increased Male Offspring in Paternal Backcross Matings," *Tissue Antigens* 15:329–332 (1980).

Chapter 4

P. 61. DANIEL NEBERT OF D. W. Nebert et al., "Genetic Differences in Mouse Cytochrome P_1-450-mediated Metabolism of Benzo(a)pyrene *in vitro* and Carcinogenic Index *in vivo*," in D. J. Jollow et al., eds., *Biological Reactive Intermediates*, New York, Plenum Press, p. 125.

P. 61. ONE STUDY, CONDUCTED G. Kellermann et al., "Aryl Hydrocarbon Hydroxylase Inducibility and Bronchogenic Carcinoma," *New England Journal of Medicine*, 289: 934–937 (1973).

P. 62. IN ONE STUDY R. Körsgaard et al., "Smoking Habits and Aryl Hydrocarbon Hydroxylase Inducibility in Patients with Malignant Tumors of the Respiratory Tract," *Cancer Letters*, 50–52 (1979).

P. 62. A FINNISH STUDY C. G. Gahmberg et al., "Induction of Aryl Hydrocarbon Hydroxylase Activity and Pulmonary Carcinoma," *International Journal of Cancer* 23: 302–305 (1979).

P. 62. RICHARD KOURI OF   Richard Kouri, personal communication.

P. 63. IN ONE OF   R. Kouri and D. Nebert, "Genetic Regulation of Susceptibility to Polycyclic-Hydrocarbon-Induced Tumors in the Mouse," *Origins of Human Cancer*, New York, Cold Spring Harbor Laboratory, 1977, pp. 811–835.

P. 64. PERHAPS THE MOST   D. W. Nebert et al., "Birth Defects and Aplastic Anemia: Differences in Polycyclic Hydrocarbon Toxicity Associated with the *Ah* Locus," *Archives of Toxicology* 39: 109–132 (1977).

P. 66. THE CHROMOSOMAL DEFECT   U. Francke et al., "Aniridia-Wilms' Tumor Association: Evidence for Specific Deletion of 11p13," *Cytogenetics and Cell Genetics* 24: 185–192 (1979); H. Hashem and S. Khalifa, "Retinoblastoma: A Model of Hereditary Fragile Chromosome Regions," *Human Heredity* 25: 35–49 (1975).

P. 68. IN ONE GROUP   Edward J. Calabrese, *Pollutants and High-Risk Groups*, New York, John Wiley, 1979, p. 19.

P. 69. NICHOLAS PETRAKIS, A   Nicholas Petrakis, personal communication; N. L. Petrakis, "Breast Secretory Activity and Breast Cancer Epidemiology," in J. J. Mulvihill et al., eds., *Genetics of Human Cancer*, New York, Raven Press, 1977, pp. 297–299.

P. 70. ELEVEN FAMILIES, WITH   M-C. King et al., "Allele Increasing Susceptibility to Human Breast Cancer May Be Linked to the Glutamate-Pyruvate Transaminase Locus," *Science* 208: 406–408 (1980).

P. 71. RECENTLY, HOWEVER, ANOTHER   Z. P. Harsanyi et al., "Mutagenicity of Melanin from Human Red Hair," *Experimentia* 36: 291–292 (1980).

P. 72. ONE STUDY SHOWED   H. T. Lynch et al., "Management and Control of Familial Cancer," in J. J. Mulvihill et al., *Genetics of Human Cancer*, New York, Raven Press, 1977, pp. 239–241.

P. 72. WITH THAT IN   F. Hecht and B. K. McCaw, "Chromosome Instability Syndromes," in *ibid.*, p. 114.

P. 72. THE SEARCH FOR   *Time*, September 24, 1979, p. 77; A. H. Cohen et al., "Hereditary Renal-Cell Carcinoma Associated with Chromosomal Translocation," *New England Journal of Medicine 301:* 592 (1979).

P. 75. SEVEN YEARS LATER   J. J. Mulvihill, in "Individual Differences in Cancer Susceptibility," (NIH Conference), *Annals of Internal Medicine* 92: 813 (1980).

P. 76. ONE OF THE   P. N. Cunliffe et al., "Radiosensitivity in A-T," *British Journal of Radiology* 48: 374–376 (1975).

P. 76. MICHAEL SWIFT, OF   Michael Swift, personal communication.

P. 76. THE TWO GROUPS   N. Torben Bech-Hansen et al., "The Association of *in vitro* Radiosensitivity and Cancer in a Family with Acute Myelogenous Leukemia," unpub. preprint, 1981; P. J. Smith and M. C. Patterson, "Abnormal Responses to Mid-Ultraviolet Light of Cultured Fibroblasts from Patients with Disorders Featuring Sunlight Sensitivity," *Cancer Research* 41: 511–518 (1981).

P. 77. BECAUSE SOME PEOPLE   Gina Bari Kolata, "Testing for Cancer Risk," *Science* 207: 967–969 (1980); personal communication.

Chapter 5

P. 79. ONE CASUAL SURVEY   Unpublished survey performed as a class exercise in Microbiology course, Cornell Medical Center, New York, N.Y.

P. 80. NEXT CAME COMPANIES   T. H. Maugh, II, "Carcinogens in the Workplace: Where to Start Cleaning Up," *Science* 197: 1268–1269 (1977).

P. 82. HE WENT TO   G. M. Lower, Jr. et al., "N-Acetyltransferase Phenotype and Risk in Urinary Bladder Cancer: Approaches in Molecular Epidemiology. Preliminary Results in Sweden and Denmark," *Environmental Health Perspectives* 29: 71–79 (1979).

P. 84. STUDIES IN COTTON   H. E. Evans and N. Bognacki, "α_1-Antitrypsin Deficiency and Susceptibility to Lung Disease," *Environmental Health Perspectives* 29: 57–61 (1979).

P. 84. AS FAR BACK   S. Lakshminarayan, "Diagnosis and Management of Hypersensitivity Pneumonitis," *Practical Cardiology* 6: 45 (1980).

P. 85. "PERSONALLY, IF I . . ."   Hugh Evans, personal communication.

P. 85. BETWEEN MARCH 1968   S. R. Jones et al., "Sudden Death in Sickle Cell Trait," *New England Journal of Medicine* 22: 323–325 (1970).

P. 87. THE DUPONT COMPANY   Editorial, *Washington Post*, March 26, 1980.

P. 87. ACCORDING TO CHARLES   Charles Reinhardt, personal communication.

P. 88. RECENTLY, THREE   E. J. Calabrese et al., "Effects of Environmental Oxidant Stressors on Individuals with a G-6-PD Deficiency with Particular Reference to an Animal Model," *Environmental Health Perspectives* 29: 49–55 (1979).

P. 88. COPPER'S ABILITY TO   National Academy of Sciences, *Drinking Water and Health*, Washington, D.C., National Academy of Sciences, 1977.

P. 89. IN OCTOBER 1978   *New York Times*, February 2, 1980; *Chemical and Engineering News*, October 22, 1979, pp. 7–8.

P. 90. LIGHT-SKINNED PEOPLE   E. Shmuness, "The Importance of Pre-Employment Examination in the Prevention and Control of Occupational Skin Disease," *Journal of Occupational Medicine* 22: 407–409.

P. 91. "IT MAKES NO . . ."   Editorial, *Chemical Week*, August 13, 1980, p. 5.

Chapter 6

P. 95. MARIO DE LUISE   M. De Luise et al., "Reduced Activity of the Red-Cell Sodium-Potassium Pump in Human Obesity," *New England Journal of Medicine*, 303: 1017–1022 (1980).

P. 98. IN 1971, JEROME ROTTER   J. I. Rotter et al., "Genetic Heterogeneity of Hyperpepsinogenemic I and Normopepsinogenemic I Duodenal Ulcer Disease," *Annals of Internal Medicine*, 91: 372–377 (1979).

P. 100. IN 1980, THEY   A. Nomura et al., "Serum Pepsinogen I as a Predictor of Stomach Cancer," *Annals of Internal Medicine* 93: 537–540 (1980).

P. 101. THE DISORDER STRIKES   R. Sittoun et al., "HL-A and Pernicious Anemia," *New England Journal of Medicine* 293: 1324 (1975).

P. 101. CHOCOLATES, WHICH CONTAIN   M. Sandler et al., "A Phenylethylamine Oxidizing Defect in Migraine," *Nature* 250: 335–337 (1974).

P. 101. NEARLY ALL CHEESES   M. B. H. Youdin et al., "Conjugation Defect in Tyramine-Sensitive Migraine," *Nature* 230: 127–128 (1971).

P. 101. VEGETABLES LIKE CABBAGE   G. R. Fraser, "The Genetics of Thyroid Disease," in A. G. Steingberg and A. C. Bearn, eds., *Progress in Medical Genetics,* Vol. XI, New York, Grune & Stratton, 1969, pp. 89–115; T. H. Shepard and S. M. Gartler, "Increased Incidence of Non-tasters of Phenylthiocarbamide among Congenital Athyreatic Cretins," *Science* 131: 929 (1966).

P. 102. KING GEORGE III   Ida Macalpine and Richard Hursten, "Porphyria and King George III," *Scientific American,* July 1969, pp.: 38–46.

P. 103. STUDIES OF IDENTICAL   S. A. Atlas and D. W. Nebert, "Pharmacogenetics and Human Disease," in D. W. Park and R. L. Smith, eds., *Drug Metabolism—from Microbe to Man,* London, Taylor & Francis, 1976, pp. 393–430.

P. 104. IN 1969, PHYSICIANS   G.P. Lewis et al., "The Role of Genetic Factors and Serum Protein Binding in Determining Drug Response as Revealed by Comprehensive Drug Surveillance," *Annals of the New York Academy of Science* 179: 729 (1971).

P. 105. IN 1952, TWO   F. T. Evans et al., "Sensitivity to Succinylcholine in Relation to Serum-Cholinesterase," *Lancet* June 21, 1952, pp. 1229–1230.

P. 105. HITOSHI SHICHI OF   H. Shichi et al., "The *Ah* Locus: Genetic Differences in Susceptibility to Cataracts Induced by Acetaminophen," *Science* 200: 539 (1978).

P. 106. CATARACTS CAN BE   R. E. Kouri and D. W. Nebert, "Genetic Regulation of Susceptibility to Polycyclic-Hydrocarbon-Induced Tumors in the Mouse," *Origins of Human Cancer,* Cold Spring Harbor Laboratory, 1970, p. 811.

P. 106. THE ACETYLATOR TYPE   H. M. Perry, Jr., et al., "Relationship of Acetyl Transferase Activity to Antinuclear Antibodies and Toxic Symptoms in Hypertensive Patients Treated with Hydralazine," *Journal of Laboratory and Clinical Medicine* 76: 114 (1970); H. B. Hughes et al., "Metabolism of Isoniazid in Man as Related to the Occurrence of Peripheral Neuritis," *American Review of Tuberculosis* 70: 266 (1954).

P. 106. APLASTIC ANEMIA, A   A. A. Yunis, "Chloramphenicol-Induced Bone Marrow Suppression," *Seminars in Hematology* 10: 225 (1973).

P. 106. TOXIC REACTIONS TO   P. H. Wooley et al., "HLA-DR Antigens and Toxic Reaction to Sodium Aurothiomalase and D-Penicillamine in Patients with Rheumatoid Arthritis," *New England Journal of Medicine* 303: 300–301 (1980).

P. 106. WARFARIN AND DICUMAROL   R. A. O'Reilly et al., "Hereditary Transmission of Exceptional Resistance to Coumarin Anticoagulant Drugs: The First Reported Kindred," *New England Journal of Medicine* 271: 809 (1964).

P. 107. SEVERE HEMOLYTIC ANEMIA   S. A. Atlas and D. W. Nebert, "Pharmacogenetics and Human Disease."

Chapter 7

P. 109. A STUDY OF   N. E. Morton et al., "An Estimate of the Mutational Damage on Man from Data on Consanguineous Marriages," *Proceedings of the National Academy of Science U.S.A.* 42: 855–863 (1956).

P. 111. AMONG THE CANCERS   G. N. Rogentine et al., "Prolonged Disease-Free Survival in Bronchogenic Carcinoma Associated with HLA-Aw19 and B5. A Follow-up," in *HLA and Disease*, Paris, INSERM, 1976, p. 234.

P. 112. AT THE BEGINNING   E. B. Ford, *Ecological Genetics*, New York, John Wiley, 1975.

P. 114. ACCORDING TO THE   *Morbidity and Mortality Weekly Report*, Center for Disease Control, Atlanta, Georgia, October 10, 1980.

P. 114. THE SIGNIFICANCE OF   L. H. Miller et al., "Erythrocyte Receptors for (Plasmodium knowlesi) Malaria: Duffy Blood Group Determinants," *Science* 189: 561–563 (1975).

P. 115. THE LABORATORY WAS   L. H. Miller et al., "The Resistance Factor to Plasmodium vivax in Blacks: The Duffy-Blood-Group Genotype, FyFy," *New England Journal of Medicine* 295: 302–304 (1976); Louis Miller, personal communication.

P. 116. AND SCIENTISTS IN   L. Luzzatto et al., "Glucose-6-Phosphate Dehydrogenase Deficient Red Cell: Resistance to Infection by Malarial Parasites," *Science* 164: 839–841 (1969).

P. 116. IN SARDINIA, THE   M. Siniscalco et al., "Population Genetics of Haemoglobin Variants, Thalassaemia and Glucose-6-Phosphate Dehydrogenase Deficiency, with Particular Reference to the Malaria Hypothesis," *Bulletin of the World Health Organization* 34: 379–393 (1966).

P. 119. UNTIL PABLO RUBINSTEIN   P. Rubinstein et al., "A Recessive Gene Closely Linked to HLA-D and with 50 Percent Penetrance," *New England Journal of Medicine*, 297: 1036–1040 (1977).

P. 119. FINALLY IN 1979   J-W Yoon et al., "Isolation of a Virus from the Pancreas of a Child with Diabetic Ketoacidosis," *New England Journal of Medicine* 300: 1173–1179 (1979).

P. 119. IN 1979, COLEMAN   In *The Sciences* July/Aug., 1979, p. 5; D. Coleman, "Obesity Genes. Beneficial Effects in Heterozygous Mice," *Science* 203: 663 (1979).

P. 121. FRED BERGMAN OF   Fred Bergman, personal communication.

Chapter 8

P. 126. ONE GROUP OF SCIENTISTS   Quoted in J. Beckwith, "Social and Political Uses of Genetics in the United States: Past and Present," in M. Lappé and R. S. Morison, eds., *Ethical and Social Issues Posed by Human Uses of Molecular Genetics*, New York, New York Academy of Sciences, 1976, p. 47.

P. 130. IN SWITZERLAND, FOR EXAMPLE   F. Vogel and A. G. Motulsky, *Human Genetics*, New York, Springer-Verlag, 1979, pp. 182–184.

P. 134. THEN ASBJORN FÖLLING   *Genetic Screening: Programs, Principles, and Research*, Washington D.C., National Academy of Science, 1975.

P. 134. FIVE YEARS AFTER   *ibid*.

P. 136. HONEY BEES, IT SEEMS   Donald R. Griffin, "Responsiveness and Awareness of Animals," Presidential Lecture, Ninth Annual Meeting of the Society for the Neurosciences, November 2, 1979; Donald Griffin, personal communication.

P. 140. BUT, IN 1937   A whole series of articles has been written on the studies of twins reared apart, most of which touch on the development of the discipline: Constance Holden, "Identical Twins Reared Apart," *Science*, 207: 1323–1327 (1980); Edwin Chen, "Twins Reared Apart: A Living Lab," *New York Times Magazine*, December 9, 1979, p. 112; Peter Watson, "Uncanny Twins," *Sunday Times Weekly Review*, May 25, 1980, and many others.

P. 140. IN 1979, A BREAK CAME   *ibid*.

P. 142. NOTED BOUCHARD: Thomas Bouchard, personal communication.

P. 146. "LOOKING ACROSS THE TWINS' . . ."   T. Bouchard et al., "The Minnesota Study of Twins Reared Apart: Project Description and Sample Results in the Developmental Domain," unpublished paper.

P. 146. ONE STUDY, PERFORMED BY   Described in Patricia McBroom, *Behavioral Genetics*, Monographs 2, Bethesda, Md., National Institute of Mental Health 1980, pp. 136–138.

P. 147. ANOTHER RESEARCHER, RONALD WILSON   R. Wilson, "Mental and Motor Development in Infant Twins," *Developmental Psychology* 7: 3 (1972).

P. 147. WILSON'S SECOND STUDY   R. Wilson, "Twins: Early Mental Development," *Science* 175: 914–917 (1972); R. Wilson, "Synchronies in Mental Development: An Epigenetic Perspective," *Science* 202: 939–947 (1978).

Chapter 9

P. 150. THE RESEARCHERS WHO   S. Kety and P. H. Wender, "Psychiatric Genetics: Studies of Adoptees and Their Families," Annual Meeting of the Association for Research in Nervous and Mental Disease, New York, December 5–6, 1980; S. Kety, personal communication.

P. 155. ONE SUCH STUDY   M. Schuckit and V. Rayses, "Ethanol Ingestion: Differences in Blood Acetaldehyde Concentration in Relatives of Alcoholics and Controls," *Science* 203: 54–59 (1979).

P. 156. SOON THEY FOUND ONE   B. S. Centervall and M. H. Criqui, "Prevention of the Wernicke-Korsakoff Syndrome," *New England Journal of Medicine* 299: 285–289 (1978); J. P. Blass and G. E. Gibson, "Abnormality of a Thiamine-Requiring Enzyme in Patients with Wernicke-Korsakoff Syndrome," *New England Journal of Medicine* 297: 1367–1370 (1977).

P. 161. ONE OF THESE ENZYMES   In Patricia McBroom, *Behavioral Genetics*, pp. 50–54; conflicting evidence comes from, among others, Elliot Gershon, "Genetics of Major Psychoses," Annual Meeting of the Association for Research in Nervous and Mental Disease, New York, December 5–6, 1980.

P. 162. IN THE MID-1970s   D. Comings, "Pc 1 Duarte, a Common Polymorphism of a Human Brain Protein, and its Relationship to Depressive Disease and Multiple Sclerosis," *Nature* 277: 28–32, (1979); D. Comings, personal communication.

P. 166. IN THE LATE 1960s   D. Roulland-Dussoix and H. W. Boyer, "The Escherichia coli B Restriction Endonuclease," *Biochimica et Biophysica Acta* (Amsterdam) 195: 219–229 (1969).

P. 168. ONE PROMINENT SCIENTIST   Park Gerald, speech at the Annual Meeting of the Association for Research in Nervous and Mental Disease, New York, December 5–6, 1980.

Chapter 10

P. 171. WHEN ASKED WHY   Interview with William Shockley, *Playboy*, August 1980, p. 740.

P. 173. "INTELLIGENCE IS WHAT . . ."   The phrase is practically a cliché in psychology, used generally to point up the difficulty of defining exactly what intelligence is and what IQ tests measure.

P. 173. "FACULTY OF THOUGHT . . ."   Quoted in J.C. de Fries et al., "Genetics of Special Cognitive Abilities," *Annual Review of Genetics* 16: 179–207 (1972).

P. 173. "CAPACITY FOR REASONING . . ."   *Random House Dictionary of the English Language*, New York, Random House, 1973, p. 739.

P. 174. THIS CONTENTION RECENTLY   In Patricia McBroom, *Behavioral Genetics*, pp. 177–178.

P. 175. THE "FRAGILE X" CHROMOSOME WAS   " 'Fragile X' Solves Mystery of Excess Male Retardates," *Medical World News*, 20 August 1979, pp. 27–29.

P. 176. RECENT TESTS OF THE IQS   O. Thalhammer et al., "Intellectual Level (IQ) in Heterozygotes for Phenylketonuria (PKU)," *Human Genetics* 38: 285–288 (1977).

P. 177. BUT A RELIABLE BLOOD TEST   C. R. Scriver and C. L. Crow, "Phenylketonuria: Epitome of Human Biochemical Genetics," *New England Journal of Medicine* 303: 1394–1400 (1980).

P. 177. TESTS OF THE BLOOD GROUPS   J. B. Gibson et al., "IQ and ABO Blood Groups," *Nature* 246: 498–499 (1973).

P. 178. ARTHUR JENSEN, WHO RECENTLY   Arthur Jensen, *Bias in Mental Testing*, New York, Free Press, 1980.

P. 178. PERHAPS THE MOST TELLING   Letter sent to members of the Genetics Society of America by the editorial board of *Genetics*, 1976.

P. 179. SUPPORT FOR THIS THESIS   In P. McBroom, *Behavioral Genetics* pp. 132–133.

P. 181. ABOUT A DECADE AGO   A. L. Gary and J. Glover, *Eye Color, Sex, and Children's Behavior*, Chicago, Nelson-Hall, 1976.

P. 181. SOON OTHER RESEARCHERS   Peter W. Post, personal communication.

P. 183. IN 1968, TWO BOSTON-BASED   S. Walzer and P. Gerald, "Social Class and Frequency of XYY and XXY," *Science* 190: 1228–1229 (1975); "XYY: Harvard Researcher Under Fire Stops Newborn Screening," *Science* 188: 1284–1285 (1975).

P. 185. IN 1976, A STUDY   H. A. Witkin et al., "Criminality in XYY and XXY Men," *Science* 193: 547–555 (1976).

Chapter 11

P. 188. THE TINY FARMING VILLAGE   G. Stamatoyannopoulos, "Problems of Screening and Counseling in the Hemoglobinopathies," in A. G. Motulsky and W. Lenz, eds., *Birth Defects: Proceedings of the Fourth International Conference*, Armsterdam, Excerpta Medica, 1974, pp. 268–275; R. H. Kenen and R. M. Schmidt, "Stigmatization of Carrier Status: Social Implications of Heterozygote Genetic Screening Programs," *American Journal of Public Health* 68(11): 1116–1120 (1978).

P. 192. THE POTENTIAL OF MASS SCREENING   P. Reilly, "Government Support of Genetic Screening," *Social Biology*, 25: 23–32 (1978).

P. 197. IN DECEMBER 1971   An excellent short account of the history of government's dealings with genetic screening appears in Marc Lappé, *Genetic Politics: The Limits of Biological Control*, New York, Simon and Schuster, 1979.

P. 198. THE GUIDELINES OF LAPPÉ'S   *ibid.*, pp. 102–104.

P. 199. ON THE OTHER SIDE Editorial, *Medical World News*, November 28, 1979.

P. 203. IN 1973, MICHAEL SWIFT M. Swift et al., *Cancer Research* 36: 209 (1976); Gina Bari Kolata, "Testing for Cancer Risk," *Science* 207: 967–971 (1980); M. Swift, "Cancer Risk Counseling," *Science* 210: 1074 (1980); M. Swift, personal communication.

P. 204. A STUDY IN ENGLAND In M. Lappé, *Genetic Politics*, p. 51.

P. 205. AS MICHAEL SWIFT M. Swift, personal communication.

P. 206. IN 1974, EDWIN NAYLOR E. W. Naylor, "Genetic Screening and Genetic Counseling: Knowledge, Attitudes, and Practices in Two Groups of Family Planning Professionals," *Social Biology* 22: 304–314 (1975).

P. 206. THE SURVEY ASKED *ibid.*, p. 307.

P. 210. AS JAMES BOWMAN HAS NOTED Quoted in *Genetics and the Quality of Life*, Pergamon Press, 1975, pp. 149–150.

INDEX

ABOUT THE AUTHORS

DR. ZSOLT HARSANYI, Associate Professor of Genetics at Cornell Medical College in New York, is also director of the committee advising the U.S. Congress on the application of genetic research to human life.

RICHARD HUTTON, a New Yorker, writes regularly on science for leading magazines and is the author of four previous scientifically oriented books.

New from

BANTAM NEW AGE ❦ BOOKS

A Search for Meaning, Growth and Change

☐ **NEW RULES:** *Searching for Self-Fulfillment in a World Turned Upside Down by Daniel Yankelovich (#22511-1 · $3.95)*

A new American philosophy is emerging, a genuine cultural revolution whose ultimate goal may be to humanize our industrial society. Daniel Yankelovich, one of the country's most respected analysts of social trends and public attitudes, has synthesized forty years of research and hundreds of personal interviews to arrive at the stunning and hopeful conclusions in this book.

☐ **GENETIC PROPHESY:** *Beyond the Double Helix by Zsolt Harsanyi and Richard Hutton (#22601-7 · $3.95)*

Through the study of the intricate spirals of DNA—the essence of life itself—researchers are now unraveling once unfathomable mysteries about our vulnerability to cancer, heart disease, alcoholism and even suicide. GENETIC PROPHECY is the first full report on the Genetic Revolution—how it is predicting our health, changing our lives and shaping our future.

☐ CREATIVE VISUALIZATION
by Shakti Gawain (#22689-4 · $3.50)

Thousands of readers have already adopted this inspirational introduction and workbook to using the art of mental energy to transform and greatly improve their lives. Famous teacher Shakti Gawain provides easy-to-follow exercises, meditations, affirmations and other techniques to help you tap into the natural goodness and beauty in all that is around you.

☐ THE MIND'S I: Fantasies and Reflections of Self and Soul by Douglas Hofstadter and Daniel Dennett (A LARGE FORMAT BOOK · #01412-9 · $6.95/$7.95 in Canada)

In this unique, mind-jolting book, Douglas Hofstadter, author of the Pulitzer Prize-winning bestseller GÖDEL, ESCHER, BACH, and philosopher Daniel Dennett, author of the widely-acclaimed BRAINSTORMS, explore the meaning of self and consciousness through the perspectives of literature, thought experiments, wondrous fantasies and humorous dialogues.

Buy these books at your local bookstore or use this handy coupon for ordering:

BANTAM NEW AGE BOOKS

Bantam New Age Books are for all those interested in reflecting on life today and life as it may be in the future. This important new imprint features stimulating works in fields from biology and psychology to philosophy and the new physics.

☐	22511	**NEW RULES: SEARCHING FOR SELF-FULFILLMENT IN A WORLD TURNED UPSIDE DOWN** Daniel Yankelovich	$3.95
☐	22510	**ZEN IN THE MARTIAL ARTS** J. Hyams	$2.95
☐	20650	**STRESS AND THE ART OF BIOFEEDBACK** Barbara Brown	$3.95
☐	13578	**THE DANCING WU LI MASTERS:** An Overview of the New Physics Gary Zukav	$3.95
☐	14131	**THE FIRST THREE MINUTES** Steven Weinberg	$2.95
☐	13470	**LIFETIDE** Lyall Watson	$3.50
☐	12478	**MAGICAL CHILD** Joseph Chilton Pearce	$3.50
☐	22786	**MIND AND NATURE:** A Necessary Unity Gregory Bateson	$3.95
☐	20322	**HEALTH FOR THE WHOLE PERSON** James Gordon	$3.95
☐	20708	**ZEN/MOTORCYCLE MAINTENANCE** Robert Pirsig	$3.95
☐	20693	**THE WAY OF THE SHAMAN** Michael Hamer	$3.95
☐	10949	**TO HAVE OR TO BE** Fromm	$2.95
☐	14821	**IN THE SHINING MOUNTAINS** David Thomson	$3.95
☐	14526	**FOCUSING** Eugene Gendlin	$3.50
☐	13972	**LIVES OF A CELL** Lewis Thomas	$2.95
☐	14912	**KISS SLEEPING BEAUTY GOODBYE** M. Kolbenschlag	$3.95